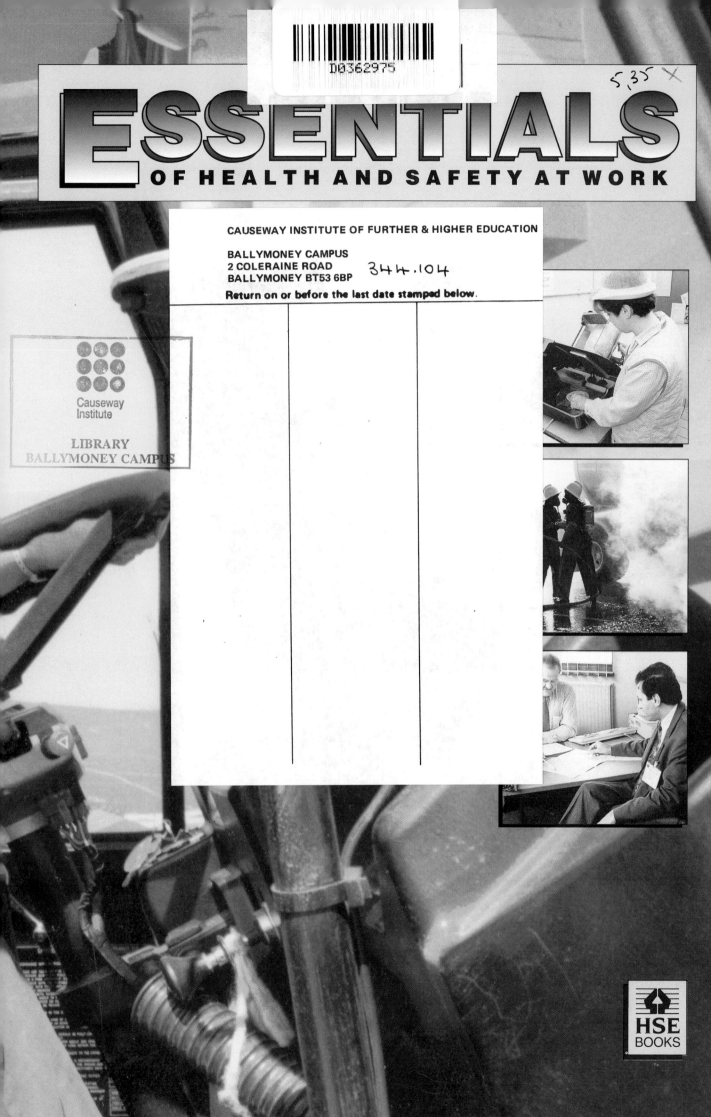

D0362975

5.35 X

ESSENTIALS
OF HEALTH AND SAFETY AT WORK

HSE
BOOKS

This guidance is issued by the Health and Safety Executive. Following the guidance is not compulsory and you are free to take other action. But if you do follow the guidance you will normally be doing enough to comply with the law. Health and safety inspectors seek to secure compliance with the law and may refer to this guidance as illustrating good practice.

CONTENTS

CONTINUED . . .

CONTENTS

PROCEDURES

PEOPLE

REFERENCES

INTRODUCTION

Every year about 500 people are killed at work and several hundred thousand more are injured and suffer ill health.

There are legal health and safety requirements that you have to meet, but accidents also cost money and time - people off work, material costs and damage to buildings, plant or product. These costs are often not covered by insurance.

If you run a small business, both time and money may be in short supply. This book will help you answer two questions:

'What are the most important things I have to do?' and 'How far do I need to go to comply with the law?'

It gives practical advice on coping with health and safety in the reality of business life.

Health and Safety at Work etc. Act 1974

1974 CHAPTER 37

Act to make further provision for securing the health, safety and welfare of persons at work, for protecting others against risks to health or safety in connection with the activities of persons at work, for controlling the keeping and use and preventing the unlawful acquisition, possession and use of dangerous substances, and for controlling certain emissions into the ...

This book, the law and guidance

 There are two main kinds of health and safety law. Some is very specific about what you must do, but some, such as the Health and Safety at Work etc Act 1974 (HSW Act), is general, requiring you to do what is 'reasonably practicable' to ensure health and safety. A good deal of guidance is published by the Health and Safety Commission (HSC) and Health and Safety Executive (HSE) to help you to decide what this means in practice.

In this book, the word 'must' indicates a definite legal requirement, eg 'You must provide enough light in a workplace or, where necessary, enough storage for clothing'.

We also give many important 'dos and don'ts' taken from recognised and generally accepted HSE guidance. These are not compulsory but they do show what we consider to be 'reasonably practicable', eg 'Do not drain petrol tanks over a pit'. If you choose to follow them, you will normally be doing enough. You are free to take other action, but you might be called on to show that you have done what is reasonably practicable to meet a good standard. Inspectors might refer to guidance to help a court decide what was needed to comply with the law.

Some 'dos and don'ts' have a special status because they are taken from Approved Codes of Practice. A court will find you at fault if you do not follow the guidance in Approved Codes unless you can show you have done something else at least as safe. Material taken from Approved Codes is in italics, eg *Do not overload floors*.

The book also gives tips on good practice which we hope you will find helpful. These are indicated by phrases such as 'Think about . . . ', eg 'Think about marking the edges of vehicle pits'.

The book is in two parts -

Chapter 1 suggests a way to tackle the basics of health and safety. It shows how to identify, assess and control the activities that might cause harm in your business.

Chapters 2 to 19 are for anyone who needs to know more about a particular subject. At the start of each chapter you are told about the important legal requirements.

Information about other publications is given in the references section at the back. You may find that this book tells you all you need to know, but if not, the references will help you find out more.

Looking at your business in the way this book suggests will help you stay safe. It will go a long way to satisfying the law - including the risk assessment that you must do under the Management of Health and Safety at Work Regulations 1992. It might save you money as well!

ORGANISING FOR SAFETY

WHAT THE LAW REQUIRES

Under the HSW Act, you have to ensure the health and safety of yourself and others who may be affected by what you do or fail to do. This includes people who: work for you, including casual workers, part-timers, trainees and sub-contractors; use workplaces you provide; are allowed to use your equipment; visit your premises; may be affected by your work, eg your neighbours or the public; use products you make, supply or import; or use your professional services, eg if you are a designer.

The Act applies to all work activities and premises and everyone at work has responsibilities under it, including the self-employed.

New regulations have replaced and updated much of the old law on health and safety, but there are specific laws applying to certain premises, such as the Factories Act 1961 and the Offices, Shops and Railway Premises Act 1963.

Don't forget your own experience. Have you had any accidents? What happened?

See the references section at the back of the book for details of publications which relate to **MANAGING HEALTH AND SAFETY**

Some basic information

You must:

❑ have a written, up to date health and safety policy if you employ five or more people

❑ carry out a risk assessment (and if you employ five or more people, record the main findings and your arrangements for health and safety)

❑ notify occupation of premises to your local inspector if you are a commercial or industrial business

❑ display a current certificate as required by the Employers' Liability (Compulsory Insurance) Act 1969 if you employ anyone

❑ display the Health and Safety Law poster for employees or give out the leaflet

❑ notify certain types of injuries, occupational diseases and events (see Chapter 16)

❑ consult union safety representatives, representatives of employee safety or employees themselves on issues such as changes affecting health and safety and the provision of information and training

❑ take account of the special needs of workers who are new or expectant mothers

❑ not employ children of under school leaving age, apart from on authorised work experience schemes, if you are an industrial undertaking.

Hazard and risk

A hazard is anything that can cause harm (eg chemicals, electricity, working from ladders etc).

Risk is the chance (big or small) of harm being done.

As an example, think about a can of solvent on a shelf. There is a hazard if the solvent is toxic or flammable, but very little risk. The risk increases when it is taken down and poured into a bucket. Harmful vapour is given off and there is a danger of spillage. If it were then used to clean the floor, the chance of harm, ie the risk, is high.

You will see the term 'risk assessment' used in regulations and guidance. Do not be put off by this phrase - it's all about doing the things described in this chapter.

Who might be harmed ?

❑ **Workers** - including those off-site

❑ **Visitors to your premises**, eg cleaners, contractors

❑ **The public**, eg when calling in to buy your products.

Look for the hazards

Walk around your workplace. Think about what could go wrong at each stage of what you do.

Here are some typical activities where accidents happen.

- receipt of raw materials,
  eg lifting, carrying

- stacking and storage,
  eg falling materials

- movement of people and materials,
  eg falls, collisions

- processing of raw materials,
  eg exposure to toxic substances

- maintenance of buildings,
  eg roof work, gutter cleaning

- maintenance of plant and machinery,
  eg lifting tackle, isolation of equipment

- using electricity,
  eg using hand tools, extension leads

- distribution of finished goods,
  eg movement of vehicles

- dealing with emergencies,
  eg spillages, fires

In a small business most accidents are caused by a few key activities. Ignore the trivial and concentrate on those that could cause serious harm. But don't just look at the obvious ones - operations such as roof work, maintenance, and transport movements (including lift trucks) cause far more deaths and serious injuries each year than many mainstream activities.

Don't forget your own experience. Have you had any accidents? What happened?

How high are the risks?

To help you decide, think about:

- what is the worst result? Is it a broken finger, someone suffering permanent lung damage or being killed? Look at the hazard summary boxes in Chapters 2 to 19 of the book

- how likely is it to happen? How often do you do the job? How close do people get to the hazard? How likely it is that something can go wrong?

- how many people could be hurt if things did go wrong? Could this include people who don't work for you?

Don't forget non-production tasks, off-site activities and work done out of normal hours.

You should now know what your main risks are.

Are the main risks under control?

You now need to see if you are taking the right precautions. You may already be doing enough, but how can you be certain?

Use Chapters 2 to 19 of this book to help you. Look at the work, talk to people and check records. Find out what actually goes on, not what you think goes on.

If you employ five or more people, write down your findings at this stage and you will have done your 'risk assessment'.

Where to get help

You probably know quite a lot already, but perhaps need some more help. Talk to your workers first, but be wary of any bad habits they have!

You must have a source of competent advice. If not already available in-house, there are many people you can turn to for help - employers' organisations, trades associations, chambers of commerce, local training organisations, TECs/LECs, local health and safety groups, trades unions, insurance companies, suppliers of plant, equipment and chemicals, and consultants.

○ suppliers provide product data sheets and instructions on using machines, tools and chemicals etc. Containers often have helpful labels

○ safety magazines have useful articles and advertise safety products and services

○ the British Safety Council (BSC), the Royal Society for the Prevention of Accidents (RoSPA) and many other independent companies and consultants run training courses - look in the telephone directory

○ HSC publishes a newsletter about new HSE publications, changes in the law and similar items of interest - the address is at the back of the book.

HSE inspects most factories, workshops and industrial premises. If your business is an office, shop, warehouse, consumer service, restaurant, hotel, or leisure and entertainment venue, your inspector will be from the local authority (district or borough council) not HSE. Their address will be in the local telephone directory. Your inspector will be able to give you advice on the law and help you with published guidance. HSE local addresses are at the back of the book.

If you need help, but don't know where to get it - ask your inspector.

Make improvements

If you find that more needs to be done, ask yourself if you can get rid of the hazard by doing the job in a different way. For example, by using a different, safer chemical, or buying materials already cut to size instead of doing it yourself.

If not, you need to think about controlling the hazard in some other way - use Chapters 2 to 19 to help you decide how.

○ choose the most important things to tackle first

○ set realistic dates for each of the improvements needed

○ don't try to do everything at once

○ remember to agree precautions with the workforce, working together to solve problems

○ don't forget that new training and information could be needed.

If you find you have quite a lot to do, think about preparing an action plan, saying what you will do and when.

Check that precautions remain in place

Remember that things change - new materials come in, machines wear out and break down and need regular maintenance, rules get broken and people don't always do as they've been told.

The only way to find out about changes like these is by checking. Don't wait until things have gone wrong, but don't try to check everything at once. Deal with a few key issues at a time, starting with the main hazards you picked out earlier.

This also lets people know that checks will be made and that you are interested in what is happening in the workplace - not just when things have gone wrong.

Don't forget maintenance. Be guided by makers' recommendations when working out your own maintenance schedules for such items as vehicles, lift trucks, ventilation plant, ladders, portable electrical equipment, protective clothing and equipment, and machine guards.

Remember - checks are not a substitute for maintenance.

Inspectors and the law

 Health and safety laws which apply to your business are enforced by an inspector either from HSE or from your local council. Their job is to see how well you are dealing with your workplace hazards, especially the more serious ones which could lead to injuries or ill health. They may wish to investigate an accident or a complaint.

Inspectors do visit workplaces without notice but you are entitled to see their identification before letting them in.

Don't forget that they are there to give help and advice, particularly to smaller firms who may not have a lot of knowledge. When they do find problems they will aim to deal with you in a reasonable and fair way. If you are not satisfied with the way you have been treated, take the matter up with the inspector's manager, whose name is on all letters from HSE. Your complaint will certainly be investigated, and you will be told what is to be done to put things right if a fault is found.

Inspectors do have wide powers which include the right of entry to your premises, the right to talk to employees and safety representatives and to take photographs and samples. They are entitled to your co-operation and answers to questions.

If there is a problem they have the right to issue a notice requiring improvements to be made, or (where a risk of serious personal injury exists) one which stops a process or the use of dangerous equipment. If you receive an improvement or prohibition notice you have the right to appeal to an industrial tribunal.

Inspectors do have the power to prosecute a business or, under certain circumstances, an individual for breaking health and safety law, but they will take your attitude and safety record into account.

How often to check

Some important items could need looking at daily, while others can safely be left for much longer. It is for you to decide, except where the law requires some inspections or examinations to be carried out by specially appointed people. Here are some key examples:

❍ ventilation plant must be examined and tested every 14 months

❍ power press guards must be inspected at each shift

❍ scaffolds must be inspected weekly

❍ rescue equipment must be examined monthly.

Other periodic tests and examinations must be carried out by a 'competent person'.

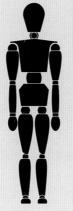

Who is a competent person?

Someone who has the necessary technical expertise, training and experience to carry out the examination or test. This could be an outside organisation such as an insurance company or other inspecting organisation, a self-employed person or one of your own staff who is capable of doing the task.

So remember, act now to get control - don't react to an accident tomorrow.

PREMISES

WHAT ARE THE HAZARDS?

There can be many dangers at work. Safety hazards include slips, trips and falls, and fire. Health hazards include poor seating, lighting and ventilation.

Assess your own working environment by using the guidelines in this chapter. Don't forget any people with disabilities who may need things like special toilet and washing facilities, wide doorways and gangways.

The law

Look at the Workplace (Health, Safety and Welfare) Regulations 1992 for the full requirements. The Regulations originally applied in full only to new and modified workplaces, but existing workplaces have been covered from 1 January 1996, replacing similar requirements, previously in the Factories Act 1961 and the Offices, Shops and Railway Premises Act 1963.

See the references section at the back of the book for details of publications which relate to **THE WORKPLACE**

A safe place of work

You must have:

❍ buildings in good repair

❍ precautions where people or materials might fall from open edges, eg fencing or guard-rails

❍ floor openings, eg vehicle examination pits, fenced or covered when not in use

❍ space for safe movement and access, eg to machinery

❍ *safe glazing, if necessary, (eg protected, toughened or thick) which is marked to make it easy to see*

❍ floors, corridors and stairs etc free of obstructions, eg trailing cables

❍ good drainage in wet processes

❍ windows that can be opened and cleaned safely. *They should be designed to stop people falling out or bumping into them when open. You may need to fit anchor points if window cleaners have to use harnesses*

❍ weather protection for outdoor workplaces, if practical

❍ outdoor routes kept safe during icy conditions, eg salted/ sanded and swept.

Also think about:

❍ machinery and furniture being sited so that sharp corners do not stick out

❍ *not overloading floors*

❍ space for storing tools and materials

❍ marking the edges of openings like vehicle pits.

Lighting

You must provide:

❍ good light - use natural light where possible *but try to avoid glare*

❍ a good level of local lighting at work stations where necessary

❍ suitable forms of lighting. Some fluorescent tubes flicker and can be dangerous with some rotating machinery (because the rotating part can appear to have stopped)

❍ special fittings for flammable or explosive atmospheres, eg from paint spraying.

Think about:

❍ light coloured walls to improve brightness (but darker colours to reduce arc-welding flash).

Moving around the premises

You must have:

❍ safe passage for pedestrians and vehicles - *you may need separate routes*

❍ level, even surfaces without holes or broken boards

❍ hand-rails on stairs *and ramps where necessary*

❍ safe doors, eg vision panels in swing doors, sensitive edges on power doors

❍ surfaces which are not slippery

❍ *well-lit outside areas - this will help security.*

Think about:

❍ marking steps, kerbs and fixed obstacles, eg by black and yellow diagonal stripes.

Designing work stations

Work stations and seating must fit the worker and the work.

Make sure that:

❍ *back rests support the small of the back and you must provide foot rests if necessary*

❍ *work surfaces are at a sensible height*

❍ *there is easy access to controls on equipment.*

Think about:

❍ well-designed tools to reduce hand or forearm injury from repeated awkward movements

❍ reducing exposure to hazardous substances, noise, heat or cold, eg local exhaust ventilation, engineering controls - there is more about these in later chapters.

Cleanliness

You must:

❍ provide clean floors and stairs, which are drained and not slippery

❍ provide clean premises, furniture and fittings (eg lights)

❍ provide containers for waste materials

❍ remove dirt, refuse and trade waste regularly

❍ clear up spillages promptly

❍ keep internal walls or ceilings clean. *They may need painting to help easy cleaning.*

Hygiene and welfare

You must provide:

○ clean, well-ventilated toilets (separate for men and women unless each convenience has its own lockable door)

- ○ wash basins with hot and cold (or warm) running water

- ○ showers for dirty work or emergencies

○ soap and towels (or a hand dryer)

○ skin cleansers, with nail brushes where necessary

○ barrier cream and skin conditioning cream where necessary

○ special hygiene precautions where necessary, eg where food is handled or prepared (the Food Hygiene (General) Regulations 1970)

○ drying facilities for wet clothes

○ certain facilities for workers working away from base

○ lockers or hanging space for clothing

○ changing facilities where special clothing is worn

○ a clean drinking water supply (*marked if necessary to distinguish it from the non-drinkable supply*)

○ rest facilities, including facilities for eating food which would otherwise become contaminated

○ *arrangements to protect non-smokers from discomfort caused by tobacco smoke in any separate rest areas, eg provide separate areas or rooms for smokers and non-smokers or prohibit smoking in rest areas and rest rooms*

○ rest facilities for pregnant women and nursing mothers.

Display screen equipment

Workers using visual display units (VDUs) need well-designed work areas with suitable lighting and comfortable, adjustable seating. This helps to reduce eye strain, headaches and back or upper limb problems. No special precautions are necessary against radiation.

For habitual users you must:

○ assess display screen work stations and reduce risks

○ plan so there are breaks or changes of activity

○ train and inform display screen users about the health and safety aspects of their work

○ provide eye tests for users on request and at regular intervals afterwards, and special spectacles where required.

Look at the Health and Safety (Display Screen Equipment) Regulations 1992 for more details.

Fire precautions

Your local fire authority rather than HSE will give you advice on this.

- ○ provide enough exits for everyone to get out easily

- ○ provide fire doors and escape routes which are clearly marked and unobstructed

- ○ provide fire escape doors which can be opened easily from the inside whenever anyone is on the premises - don't forget 'out of hours' working

- ○ never wedge open fire doors - they are there to stop smoke and flames spreading

- ○ if a wall is meant to be 'fire resisting', stop up any holes (eg around pipework) and make sure the wall continues above a 'false ceiling'

- ○ if you have a fire alarm, check regularly that it is working. Can it be heard everywhere over normal background noise?

- ○ provide enough fire extinguishers of the right type (and properly serviced) to deal promptly with small outbreaks

- ○ everyone should know what to do in case of fire. Display clear instructions and have a fire drill periodically

- ○ do people know how to raise the alarm and use the extinguishers?

- ○ call the fire brigade to any suspected outbreak of fire

- ○ you may need a fire certificate for the building - this will depend on the kind of business you run and the number of people employed in your building. The main law is the Fire Precautions Act 1971. Ask your local fire authority about this if in doubt.

- ○ if you do not need a fire certificate you will need to carry out a fire risk assessment (Fire Precautions (Workplace) Regulations 1997).

See Chapter 14 for advice on fire hazards.

Comfortable conditions

You must provide:

- ○ a reasonable working temperature in workrooms - *usually at least 16°C, or 13°C for strenuous work*

- ○ *local heating or cooling where a comfortable temperature cannot be maintained throughout each work room (eg hot and cold processes)*

- ○ *thermal clothing and rest facilities where necessary, eg for 'hot work' or cold stores*

- ○ *good ventilation - avoid draughts*

- ○ heating systems which do not give off dangerous or offensive levels of fume into the workplace

- ○ sufficient space in work rooms.

Remember that noise can be a nuisance as well as damaging to health (see Chapter 9).

PREMISES

WHAT ARE THE HAZARDS?

Falling from heights is often the worst hazard when building or doing maintenance work on existing buildings.

People may fall from roof edges or through fragile roof materials; from scaffolds, if guard-rails are not provided; from ladders, usually by overreaching or because the ladder slips; and through holes in floors and platforms, if not covered or fenced.

Other hazards include being struck by falling materials; contact with electricity (see Chapter 12); exposure to harmful substances such as asbestos, paints, glues, or cleaning materials (see Chapter 13); striking buried electric cables or gas pipes; burial by excavation collapses; and using lift trucks as temporary working platforms.

The law

Look at the Workplace (Health, Safety and Welfare) Regulations 1992 and the Provision and Use of Work Equipment Regulations 1998 (PUWER) for the full requirements.

On building maintenance these requirements are fairly general, so most of this chapter is practical advice, except where the word 'must' is used. If you are a builder, look at the Construction (Health, Safety and Welfare) Regulations 1996, which apply instead of the Workplace Regulations.

The Construction (Design and Management) Regulations 1994 place additional duties on principal contractors, clients, designers and planning supervisors, but there are some exemptions for small projects.

See the references section at the back of the book for details of publications which relate to **BUILDING WORK**

Construction work

 Most activities involving structural work are subject to the various Construction Regulations which specify standards for a wide range of matters such as safe access and safe lifting. The Regulations apply to construction, structural alteration, repair, maintenance, repointing, redecorating and external cleaning, demolition, site preparation and laying of foundations.

❍ if you are a contractor and building works are expected to last six weeks or more you must notify your HSE inspector

❍ if you use a contractor you still have legal responsibility for many matters on your premises.

Before you start

Does the work have to be done at all? If it does, plan the job to remove risks. Don't take on work for which you are unprepared. You may need specialist help with, eg demolition of buildings, digging deep trenches, roofing.

Protect other people:

❍ use barriers and signs around the workplace

❍ prevent materials falling from scaffolds by enclosing with sheeting

❍ keep children out.

Ladders

❍ ladders are often used when it would be safer to use other equipment, eg mobile tower scaffolds

❍ ladders may be used for short jobs. This can still be dangerous however, and many ladder accidents happen during work lasting 30 minutes or less

❍ longer ladders are harder to handle. They flex more in use and are harder to 'foot' effectively. Do not use a ladder longer than 6 m as a workplace unless fixed or tied

❍ when choosing a ladder, you must make sure it is strong enough for the job and check that it is in good condition, eg that no rungs are cracked or missing. Do not use makeshift or home-made ladders or carry out makeshift repairs to a damaged ladder

❍ when placing the ladder, rest its foot on a firm, level surface. Do not place it on material or equipment to gain extra height. Ladders must extend at least 1 m above the landing place unless there is a suitable hand hold to provide equivalent support

❍ angle the ladder so that the bottom will not slip outwards - four units up to each one out from the base

❍ rest the top of the ladder against a solid surface. Equipment such as ladder stays can be used to spread the load if the surface is brittle

❍ ladders used for access or as a place of work should be secured or footed to prevent movement

❍ extending ladders need an overlap of at least three rungs

❍ never paint ladders - this may hide defects.

When using a ladder:

❍ do not carry heavy items or long lengths of material up it

❍ carry light tools in a shoulder bag or holster attached to a belt so that you have both hands free to hold it

❍ do not overreach.

Step ladders

 Step ladders can be easily overturned. Do not use the top step of a step ladder to work from unless it has specially designed hand holds. Do not overreach.

Mobile tower scaffolds

When erecting the tower:

❍ follow the manufacturer's instructions - do not exceed the maximum height allowed for a given base dimension. (Normally the base: height ratio is 1:3 for an untied tower.) Do not mix components from different types of scaffold

❍ it must rest on firm, level ground - securely fix any wheels to the scaffold

❍ tie the tower rigidly to the structure if it is likely to be exposed to strong winds, used for grit blasting, or water jetting, if heavy materials are lifted up the outside of the tower, or if the base is too small for the height of the tower needed

❍ provide a safe way to get to and from the work platform, eg by an internal ladder. It is not safe to climb up the outside

❍ you must provide guard-rails and toe-boards at platforms from which someone could fall more than 2 m, or less where there is still risk of injury

❍ you must not overload the working platform. Do not apply pressure which could overturn the tower, eg by working off a ladder placed on top of the working platform. Lock any wheels and extend outriggers.

When moving the tower:

❍ check that there are no power lines in the way or obstructions, holes etc in the ground

❍ do not allow people or materials to remain on the tower

❍ beware of towers 'running away with you' when being moved down or across slopes.

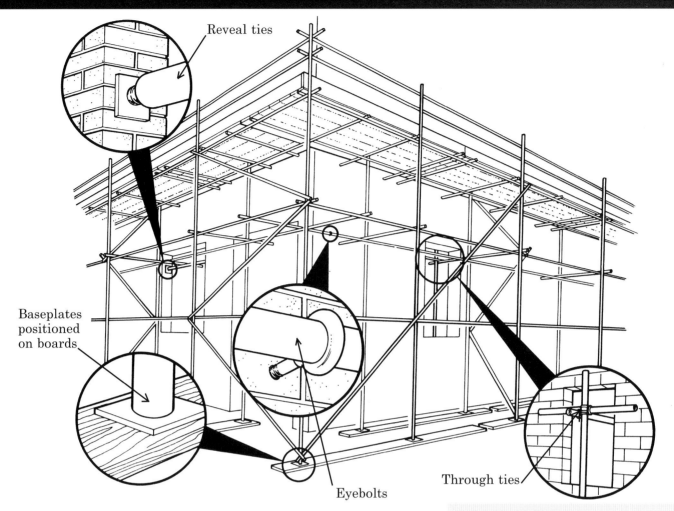

Reveal ties

Baseplates positioned on boards

Eyebolts

Through ties

General access scaffolds

When providing a scaffold you must make sure:

❍ it is erected, altered or dismantled under supervision by competent people

❍ it is based on a firm, level foundation, with vertical supports normally not more than 2-2.5 m apart

❍ it is properly tied, normally at least every 4 m vertically and 6 m horizontally, and braced

❍ platforms more than 2 m from the ground have guard-rails and toe-boards. Brick guards or similar will often be needed to provide extra protection to prevent materials falling

❍ platforms are wide enough for the work to be done there and they are fully boarded (three to five boards wide depending on use)

❍ boards are properly supported and do not overhang excessively (at least three supports not more than 1.5 m apart)

❍ there is safe ladder access onto the scaffold and between each level or lift

❍ the scaffold is inspected at least once a week, or whenever it is substantially altered or after very bad weather

❍ the person doing the inspection fully understands scaffold safety and records the results.

If you are going to work on a scaffold someone has provided for you, don't start work without checking the above points (including a look at the inspection register).

Please note: this is a stunt artist

Lift trucks

Don't use the forks of a lift truck as a workplace unless a proper working platform is fitted.

Roof work

When working on a roof you must:

- ○ have safe access onto and off the roof, eg by a general access scaffold

- ○ have safe means of moving across the roof. On sloping and fragile roofs you will need purpose-made roof ladders or crawling boards. Do not use home-made ladders or boards as these have caused many accidents. Cover openings or roof lights, or provide barriers

- ○ not walk along the line of the roof bolts above the purlins or along the roof edge of a fragile roof - this is as unsafe as walking a tightrope

- ○ use edge protection at the open edge/ eaves level of a roof to stop people and materials falling off it

- ○ not throw scaffold materials, old slates, tiles etc from the roof or scaffold where this could cause injury. Use enclosed debris chutes or lower the debris in skips or baskets.

You should:

- ○ provide permanent means of access, eg to plant rooms, ventilation equipment, skylights that need cleaning.

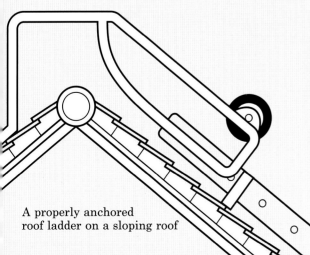

A properly anchored roof ladder on a sloping roof

Ground work

- ○ trench sides can collapse suddenly whatever the nature of the soil. Any excavation deeper than 1.2 m must have the sides sloped or supported

- ○ dig well away from underground services such as electricity cables, gas pipes etc. If you have to work near services, use service plans, locators and safe digging practice to avoid danger.

Please note: covers removed for access

Other hazards

- ○ paints, glues, cleaning materials etc used during maintenance may be a health risk - use manufacturers'/suppliers' information to identify hazards (see Chapter 13)

- ○ beware of gases found in sewers, and fumes or lack of oxygen in confined spaces (see Chapter 6).

PLANT AND MACHINERY

WHAT ARE THE HAZARDS?

Many serious accidents at work involve machinery. Hair or clothing can become entangled in rotating parts; shearing can occur between two parts moving past one another; crushing can occur between parts moving towards one another, or between machinery or parts moving towards a fixed part.

People can be struck by moving parts of machinery; cutting or severing can occur by sharp edges; and material can be ejected from machinery causing injury.

Parts of the body can be drawn in or trapped between running parts in rollers, belts and pulley drives; stabbing or puncture of the skin can occur by sharply pointed parts; and friction or abrasion is possible from contact with rough surface parts.

Extremes of temperature causing burns or scalds, and problems with electricity can also cause accidents.

See Chapter 12 for more information on the hazards of electricity.

The law

The Provision and Use of Work Equipment Regulations 1998 (PUWER) contain the requirements for work equipment to be safe. The Regulations cover maintenance and inspection, as well as guarding requirements.

Manufacturers and suppliers have duties to provide safe equipment. If in doubt, ask your inspector for more details.

See the references section at the back of the book for details of publications which relate to **MACHINERY SAFETY**

Assessing risks

Think about:

❍ all the work which has to be done on the machine in setting up, maintenance, repair, breakdowns and removing blockages, as well as normal use

❍ electrical, hydraulic or pneumatic power supplies

❍ the machine being used, not only by experienced and well-trained workers, but also by new starters, people who have changed jobs or those who have particular difficulties

❍ workers who may act foolishly or carelessly or make mistakes

❍ badly designed safeguards being inconvenient to use or easily defeated and which could encourage your workers to risk injury and break the law.

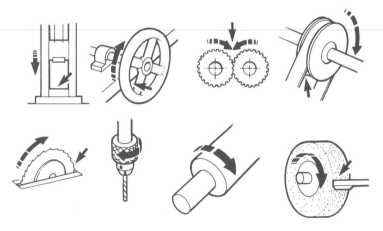

By law the supplier must provide the right safeguards, but it is also up to you to check before using any machine, and every static machine must be stable (usually fixed down).

Choose the right machine for the job and do not site machines where customers or visitors may go.

Guards

○ fixed guards enclosing the dangerous parts must be used, if practical, and must be fixed in place, eg with screws or nuts and bolts

○ think about the best materials to use - plastic may be easy to see through but can easily be damaged. Where wire mesh or similar materials are used, make sure the holes are not large enough to allow access to moving parts

○ if you have to go near dangerous parts regularly and fixed guards are not practical, you must use other methods, eg interlock the guard so that the machine cannot start before the guard is closed and cannot be opened while the machine is moving

○ in some cases, eg guillotines, trip systems such as photo-electric devices, pressure-sensitive mats or automatic guards may be used instead of fixed or interlocked guards if these are not practical

○ some machines are controlled by programmable electronic systems. Your supplier must tell you about the safety of the system. Changes to routine working or main programmes should be carried out by a competent person. Programming changes should be documented and checked

○ where guards cannot give full protection, use jigs, holders, push sticks etc if practical

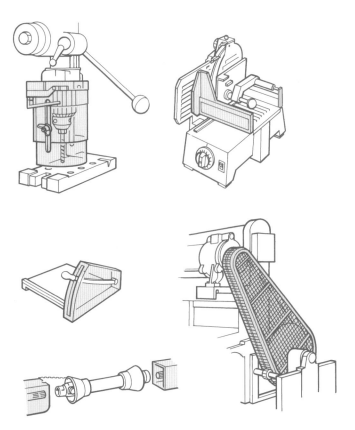

○ make sure the guards you use allow the machine to be cleaned safely

○ there is guidance on how to guard machinery such as abrasive wheels, woodworking machines, horizontal milling machines and power take-off shafts.

Machine operation

○ some workers, for example if they are young, inexperienced or have a disability, may be particularly at risk. They may need extra instruction, training or supervision. Sometimes formal qualifications are needed, for example chainsaw operators

○ you must never allow children to operate or to help at machines

○ all machine operators must be trained and given protective clothing if necessary (see Chapters 18 and 19)

○ adequate lighting must be provided for all machines.

Machinery maintenance

- you must make sure the guards and other safety devices are checked and kept in working order

- if maintenance workers need to remove guards or other safety devices, make sure that the machine is stopped first or is safe

- be alert for anyone defeating or getting around the guards or safety devices

- check the safeguarding after any modifications

- when putting in new machines think about safe maintenance.

Machine controls

You must:

- ensure control switches are clearly marked to show what they do

- have emergency stop controls if necessary, eg mushroom head push buttons within easy reach

- make sure operating controls are designed and placed to avoid accidental operation, eg by shrouding start buttons and pedals.

Operator's checklist

Check every time that:

- you know how to stop the machine before you start it

- all guards are in position and all protective devices are working

- the area around the machine is clean, tidy and free from obstruction

- you tell your supervisor at once if you think a machine is not working properly or any safeguards are faulty

- you are wearing appropriate protective clothing and equipment, such as safety glasses or shoes (see Chapter 18).

Never . . .

- use a machine unless you are authorised and trained to do so

- try to clean a machine in motion - switch it off and unplug it or lock it off

- use a machine or appliance which has a danger sign or tag attached to it. Danger signs should be removed only by an authorised person who is satisfied that the machine or process is safe

- wear dangling chains, loose clothing, gloves, rings or have long hair which could get caught up in moving parts

- distract people who are using machines.

PLANT AND MACHINERY

WHAT ARE THE HAZARDS?

There is a danger of fire and explosion from piped gas supplies and of toxic fumes (carbon monoxide) if appliances are not working properly. Explosions can occur in gas- and oil-fired plant such as ovens, stoves and boilers.

The law

The Gas Safety (Installation and Use) Regulations 1998 cover gas appliances except in factories, where the HSW Act requires the same precautions.

Gas supply

❍ if you suspect a leak you must turn off the supply and notify your gas supplier at once if gas continues to escape

❍ if in doubt, evacuate the building and inform the police as well as the gas supplier

❍ do not check for leaks with a naked flame

❍ do not turn the gas back on until the leak has been dealt with by a competent person.

Appliances

You must:

❍ use a competent fitter to install or repair your appliances

❍ not use any appliance you know or suspect is unsafe

❍ check also that the room has adequate ventilation - air inlets should not be blocked to prevent draughts, and flues and chimneys should not be obstructed

❍ get your appliances regularly serviced by a competent person.

Any business (including a self-employed person) which works on gas fittings using piped gas must be registered with the Council for Registered Gas Installers (CORGI). It is important to check this.

Plant

Explosions can be caused by ignition of unburnt fuel or flammable vapours from plant such as drying ovens on a paint spraying line. (There is a similar risk with electrically-heated equipment.)

❍ fit explosion relief and flame-failure protection as necessary

❍ interlock the heat source and the ventilation system if flammable vapours could build up to dangerous levels in the event of a ventilation failure

❍ plant, including petrol-driven compressors and liquefied petroleum gas (LPG) fuelled equipment, such as heaters and paint strippers, should be designed and operated to ensure there is enough air to burn the fuel properly

❍ there should be sufficient ventilation to remove combustion products and solvents given off

❍ make sure the operators are fully trained - use a safe procedure for purging, lighting up and shutting down the plant.

See the references section at the back of the book for details of publications which relate to **GAS- AND OIL-FIRED EQUIPMENT**

PLANT AND MACHINERY

WHAT ARE THE HAZARDS?

Maintenance is carried out to prevent problems arising and to put faults right. It may be part of a planned programme or may have to be carried out at short notice after a breakdown. It always involves non-routine activities.

Extra care is needed when getting up to and working at heights, or when doing work which requires access to unusual parts of the building (see Chapter 3).

Hazards can arise when working on machinery, including accidental/premature start-up; using hand tools and electrical equipment; during contact with materials which are normally enclosed in plant and equipment; and entering vessels or confined spaces where there may be toxic materials or a lack of air.

There may be confusion where maintenance is carried out during normal production work or poor communication where different contractors are working together at the same time on a site.

Chapter 4 provides guidance on machinery hazards, Chapter 8 on transporting and moving materials, Chapter 12 on electrical hazards, Chapter 13 on harmful substances and Chapter 14 on flammable and explosive materials.

The law
Look at the Provision and Use of Work Equipment Regulations 1998 (PUWER) and the Workplace (Health, Safety and Welfare) Regulations 1992 for the full requirements.

See the references section at the back of the book for details of publications which relate to **PLANT AND EQUIPMENT MAINTENANCE**

Before you start

Is the job necessary? Can it be done less often without increasing other risks? Should it be done by specialists? Never undertake work for which you are unprepared or not competent.

Plan the work to cut down the risks, eg the difficulties in co-ordinating maintenance and routine work can be avoided if maintenance work is performed before start-up or at shut-down periods. Access is easier if equipment is designed with maintenance in mind.

Safe working areas

❍ you must provide safe access and a safe place of work (see Chapter 3 on the use of ladders and scaffolds etc)

❍ think about the safety of maintenance workers and others who may be affected, eg other employees or contractors

❍ think about setting up signs and barriers and posting people at key points.

Safe plant

Plant and equipment must be made safe before maintenance starts.

You may need to:

- isolate electrical and other power supplies. Most maintenance should be carried out with the power off. If the work is near uninsulated overhead electrical conductors, eg close to overhead travelling cranes, the power should be cut off first

- isolate plant and pipelines containing pressurised fluid, gas, steam or hazardous material. Isolating valves should be locked off

- support parts of plant which could fall

- allow moving plant to stop

- allow components which operate at high temperatures time to cool

- place mobile plant in neutral gear, applying the brake and chocking the wheels

- clean out vessels containing flammable solids, liquids, gases or dusts and check them before hot work is carried out, to prevent explosions

- clean and check vessels containing toxic materials before work starts.

Entry into confined spaces is a high risk activity. Special precautions are necessary - see the box below.

Confined spaces

About 15 people in Britain are killed each year while working in confined spaces. Many more are seriously injured. Asphyxiation and toxic fumes are the two most common causes of death, but others include drowning in free flowing solids, eg grain silos, and fire and explosions. Two or more people are often involved in these incidents. One person is overwhelmed and then others attempt to rescue them without being adequately prepared.

What is a confined space?

Some confined spaces are fairly easy to identify, eg closed tanks, vessels and sewers. Others are less obvious but may be equally dangerous, eg open-topped tanks and vats, closed and unventilated rooms and silos. Confined spaces exist in all sectors of industry. It is not possible to provide a complete list. *You should identify where these hazards arise in your workplace and take precautions.*

Deal with confined spaces:

- identify all confined spaces

- do the work from outside where possible

- if entry is necessary follow the guidelines in Chapter 15.

Appropriate precautions will include:

- isolating the vessel to stop dust, fume or hazardous substances getting in

- cleaning to ensure fumes do not develop from residues etc while the work is being done

- testing the atmosphere to check that it is free from toxic or flammable vapours and that there is enough fresh air

- *continuing ventilation to ensure an adequate supply of fresh air*

- using non-sparking tools

- not using petrol/diesel equipment inside the confined space

- rescue equipment and trained personnel ready at hand to provide rescue and resuscitation

- safe lighting (eg low voltage).

See Chapter 18 for protective clothing and equipment and Chapter 19 for selection and training.

Clear working procedures

Use the manufacturer's maintenance instructions where provided.

Chapter 15 gives more information on safe systems, including the use of 'permits to work'.

Personal protective equipment

 Maintenance staff must be given clothing and equipment which is appropriate for the job to be done (see Chapter 18).

Hand tools

You must ensure hand tools are properly maintained, eg:

❍ **hammers** - avoid split, broken or loose shafts and worn or chipped heads. Heads should be properly secured to the shafts

❍ **files** - should have a proper handle. Never use them as levers

❍ **chisels** - the cutting edge should be sharpened to the correct angle. Do not allow the head to spread to a mushroom shape - grind off the sides regularly

❍ **screwdrivers** - should never be used as chisels, and hammers should never be used on them. Split handles are dangerous

❍ **spanners** - avoid splayed jaws. Scrap any which show signs of slipping. Have enough spanners of the right size. Do not improvise by using pipes etc as extension handles.

Vehicle repair

❍ make sure brakes are applied and wheels are chocked. Always start and run engines with brakes on and in neutral gear

❍ support vehicles on both jacks and axle stands (never rely on jacks alone)

❍ always prop raised bodies

❍ beware of the explosion risk when draining and repairing fuel tanks, and from battery gases - do not drain petrol tanks over a pit

❍ take care not to short-circuit batteries.

Use protective equipment

❍ use a tyre cage when inflating commercial tyres, particularly those with split rim wheels - explosions do happen!

❍ avoid breathing asbestos dust from brake and clutch lining pads

❍ wear protective clothing when handling battery acid.

See Chapter 18 for more information on protective clothing and equipment.

PLANT AND MACHINERY

WHAT ARE THE HAZARDS?

Devastation can occur if a piece of pressurised plant fails and bursts violently apart. There will be further risks if the system contains harmful substances such as flammable or toxic materials.

There may also be special risks associated with the maintenance of such plant (see Chapter 6).

Hazards arise with many kinds of pressurised plant and equipment including steam boilers and associated pipework; pressurised hot water boilers and heating systems; air compressors, air receivers and associated pipework; autoclaves; chemical reaction vessels; slurry tankers; and high pressure water jetting.

The law

Look at the Pressure Systems and Transportable Gas Containers Regulations 1989. The main points are summarised in this chapter.

Before you start

❍ can the job be done another way without using pressurised equipment?

❍ don't use high pressure when low pressure will do.

Good design and maintenance

❍ all plant and systems must be designed, constructed and installed to prevent danger and must have safety devices

❍ systems must be properly maintained

❍ any modifications or repairs must not give rise to danger

❍ there must be a written scheme for examination of pressure vessels, fittings and pipework, drawn up by a competent person

❍ the examinations must be carried out

❍ records must be kept.

Safe operation

You must:

❍ operate plant within the safe operating limits. Sometimes these are laid down by the manufacturer or supplier. If not, a competent person can advise you

❍ provide adequate instructions. *This should include the manufacturer's operating manual*

❍ provide instructions on what to do in an emergency.

Pressure cleaning

❍ follow the supplier's advice on high pressure jetting equipment - protective clothing and keeping other people away are important, as is electrical safety

❍ avoid using compressed air for cleaning - vacuum or low pressure nozzles can be used

❍ horseplay with compressed air is very dangerous.

See the references section at the back of the book for details of publications which relate to **PRESSURISED PLANT AND SYSTEMS**

Details of competent persons' organisations may be obtained from NSCIIB, Engineering Inspection Authorities Board, 1 Birdcage Walk, London SW1H 9JJ (Tel: 0171 222 7899)

PLANT AND MACHINERY

WHAT ARE THE HAZARDS?

Common hazards are the manual movement of loads and frequent forced or awkward movements of the body, leading, for example, to back injuries, and severe pains in the hand, wrist, arm or neck - 'repetitive strain injury'.

Moving materials mechanically is also hazardous and people can be crushed or struck by material when it falls from a lifting or moving device, or is dislodged from a storage stack.

Every year over 5000 accidents involving transport in the workplace are reported. About 70 of these accidents result in death.

People are knocked over, run over, or crushed against fixed parts by powered vehicles (eg LGVs, lift trucks and tractors) or by vehicles, plant and trailers which roll away when incorrectly parked. People also fall from vehicles - either getting on/off, working at height, or associated with loading/unloading.

This chapter is all about what you move, how and where.

The law

Look at the Workplace (Health, Safety and Welfare) Regulations 1992 and the Manual Handling Operations Regulations 1992. Requirements for mobile machinery are in the Provision and Use of Work Equipment Regulations 1998 (PUWER). See the Lifting Operations and Lifting Equipment Regulations 1998 (LOLER) for the safe use of lifting equipment.

See the references section at the back of the book for details of publications which relate to **HANDLING AND TRANSPORTING**

Before you start

Think about whether you need to do the job at all, and if there is an easier, safer way.

Manual handling

 You must avoid manual handling if a safer way (eg mechanical) is practical. Design jobs to fit the work to the person rather than the person to the work. This will take into account human capabilities and limitations and improve efficiency as well as safety.

You must:

❍ avoid manual handling where there is a risk of injury

❍ assess the risk of injury from any hazardous manual handling that can't be avoided

❍ reduce the risk of injury from hazardous manual handling.

Always consider automation or mechanisation as an alternative, but don't forget that this will introduce new hazards.

Think about:

❍ providing mechanical help such as a sack truck or hoist

❍ making the loads smaller/lighter or easier to grasp

❍ changing the system of work to reduce the effort required

❍ improving the layout of the workplace to make the work more efficient.

As a final measure, think about protective equipment, eg for hands and feet (see Chapter 18).

When lifting:

❍ stop and think - plan the lift. Do you need help? Is the area free of obstruction?

❍ place the feet - apart, leading leg forward

❍ get a firm grip - keep your arms inside the boundary formed by the legs

❍ don't jerk

❍ move the feet - don't twist the body

❍ keep close to the load;

❍ put down, then adjust.

Repetitive handling

Repeated or awkward movements which are too forceful, too fast or carried out for too long can lead to disorders of the arms, hands or neck.

Risks may arise in jobs which involve:

○ gripping, squeezing, or pressing

○ awkward hand or arm movements, eg bent wrist

○ repeated, continuous movements which are fast and unvaried or tied to the speed of a machine

○ awkward, rigid or tense body positions, eg unnatural hand positions, outstretched arms, having to lean sideways.

People are not all the same and you should take account of this when ordering tools, designing jobs, and setting work speeds.

Risks can be prevented by:

○ reducing the levels of force required, eg by maintaining equipment and by using tools with well-designed handles

○ reducing repetitive movements, eg by varying tasks, rotating jobs, using power-driven tools, reducing machine pace and introducing rest and recovery time

○ getting rid of awkward positions by changing the work station or work.

Encourage the reporting of aches and pains - these are warning signs.

Chapter 2 covers work stations and display screen equipment.

Safe lifting by machine

Safe lifting needs to be planned. Any equipment you use must have been properly designed, manufactured and tested.

Consider:

○ what you are lifting

○ its weight

○ its centre of gravity

○ how to attach it to the lifting machinery

○ who is in control of the lift

○ the safe limits of the equipment

○ rehearsing lifts if necessary.

Also . . .

○ use only certified lifting equipment, marked with its safe working load, which is not overdue for examination

○ keep the annual or six-monthly reports of thorough examination as well as any declarations of conformity or test certificates

○ never use unsuitable equipment, eg makeshift, damaged, badly worn chains shortened with knots, kinked or twisted wire ropes, frayed or rotted fibre ropes

○ never exceed the safe working load of machinery or accessories like chains, slings and grabs. Remember that the load in the legs of a sling increases as the angle between the legs increases

○ do not lift a load if you doubt its weight or the adequacy of the equipment

○ make sure the load is properly attached to the lifting equipment. If necessary, securely bind the load to prevent it slipping or falling off

○ before lifting an unbalanced load, find out its centre of gravity. Raise it a few inches off the ground and pause - there will be little harm if it drops

○ use packing to prevent sharp edges of the load from damaging slings and do not allow tackle to be damaged by being dropped, dragged from under loads or subjected to sudden loads

○ when using jib cranes, make sure any indicators for safe loads are working properly and set correctly for the job and the way the machine is configured

○ outriggers should be used where necessary

○ when using multi-slings make sure the sling angle is taken into account

○ have a responsible slinger or banksman and use a recognised signalling system.

Don't forget maintenance (see Chapter 6).

Safe stacking

Materials and objects should be stored and stacked so they are not likely to fall and cause injury.

Do . . .

❍ stack on a firm, level base. Use a properly constructed rack when needed and secure it to the floor or wall if possible

❍ use the correct container, pallet or rack for the job. Inspect these regularly for damage and reject defective ones

❍ ensure stacks are stable, eg 'key' stacked packages of a uniform size like a brick wall so that no tier is independent of another; chock pipes and drums to prevent rolling and keep heavy articles near floor level.

Do not . . .

❍ exceed the safe load of racks, shelves or floors

❍ allow items to stick out from stacks or bins into gangways

❍ climb racks to reach upper shelves - use a ladder or steps

❍ lean heavy stacks against walls

❍ de-stack by throwing down from the top or pulling out from the bottom.

Safe transport

You must:

❍ lay your workplace out so that pedestrians are safe from vehicles

❍ train your drivers.

Do . . .

❍ *separate vehicles and pedestrians where practical*

❍ mark safe crossings

❍ control pedestrian access to loading bays and delivery points

❍ ensure drivers can see clearly, and pedestrians can be seen and be aware of vehicles. Where necessary, consider the use of visibility aids, high visibility clothing, audible alarms, and lighting, both of the workplace and on vehicles

❍ make sure visiting drivers follow your rules

❍ designate level parking areas and leave vehicles in a safe state - apply the handbrake, switch off the engine and remove the keys. Chock the wheels as necessary, eg when trailers are parked overnight

❍ check vehicles daily and have faults rectified promptly

❍ supervise vehicle movements - particularly when reversing and near blind corners. Always use recognised signals

❍ load and unload materials safely, eg ensure safe access onto vehicles for loading and sheeting. Materials should be safely secured against possible movement

❍ avoid tipping on soft ground or in high winds.

Do not . . .

❍ let unauthorised people drive. Keep keys secure when vehicles are not in use

❍ let passengers ride on the vehicle unless it is designed for this.

Ionising radiations

❍ special care is needed with radioactive substances or radiation-producing equipment

❍ if IRR 85 applies to your activities, normally you must notify your inspector before starting such work

❍ permission to store and dispose of radioactive substances is given by HM Inspectorate of Pollution (HM Industrial Pollution Inspectorate in Scotland and the Environment Department of the Welsh Office). Consult also the Radioactive Substances Act 1993

❍ IRR 85 also applies where the naturally occurring gas radon is present above a defined level

❍ contractors carrying out site radiography (eg checking welds on pipework or vessels) must notify HSE before work starts

❍ smoke detectors and static eliminators often contain sources. Find out the rules for safe storage and use from the supplier and never tamper with them

❍ treat luminous articles and self-illuminating devices with similar respect.

You may need to:

❍ appoint a radiation protection adviser

❍ arrange for medical examinations/reviews and routine dose assessments of employees whom you designate as 'classified persons'

❍ appoint one or more of your employees to supervise radiation work

❍ make arrangements to cater for spills of radioactive substances, X-ray exposures failing to terminate etc

❍ get authorisation for use, storage and safe disposal of radioactive substances

❍ arrange for tests where raised levels of radon gas are likely because of workplace location, construction and ventilation, and have necessary improvements carried out.

Infra-red

❍ protective clothing may be needed to reduce warming, burning and irritation of the skin from some 'hot bodies' such as pools of molten metal. Eye protection with suitable filters should be worn to avoid discomfort and is essential with some infra-red sources, such as certain lasers (see Chapter 18).

Ultraviolet (UV)

❍ UV sources in equipment should normally be in an enclosure, or screened

❍ take care to avoid UV light, eg by wearing suitable clothing and eye protection

❍ during welding use special goggles or a face screen

❍ protect passers-by, eg with screens

❍ when fitting replacement UV lamps, choose the correct type specified by the manufacturer. Filters should be kept in place at all times and replaced after changing bulbs or if they are damaged

❍ insect killing devices with bright UV sources are often found in food premises and are not harmful to the eyes in normal use.

Lasers

❍ a laser, a concentrated beam of radiation, which may not always be visible, can be dangerous whether it is viewed directly or after reflection from a smooth surface

❍ even with low-powered lasers it is unwise to view the beam directly. Do not over-ride any interlocks

❍ maintenance workers who have to examine inside machines may be most at risk. They need to be trained and follow a work system which may include the use of eye protection

❍ high-powered lasers should normally be inside a safety interlocked enclosure - only use them after taking expert advice

❍ where lasers are used for display, eg at discotheques, there could be risks to the public - seek expert advice.

Follow good safety rules

❍ get safety information from your supplier or other specialist adviser

❍ identify, mark and (where possible) enclose sources of radiation

❍ identify and clearly mark all hazard areas

❍ maintain equipment to minimise exposure, eg by regular checking of interlocks

❍ instruct employees about dangers and precautions, including use of the correct protective equipment

❍ review procedures from time to time.

PLANT AND MACHINERY

WHAT ARE THE HAZARDS?

The three main hazards are contact with live parts, fire and explosion. Each year about 1000 accidents at work involving shock and burn are reported and about 30 of these are fatal. Fires started by poor electrical installations cause many other deaths and injuries. Explosions are caused by electrical apparatus or static electricity igniting flammable vapours or dusts.

Assess the risks from your use of electricity and use the precautions in this chapter to control them. Remember that normal mains voltage (240 volts AC) can kill. The risks are greatest when electricity is used in harsh conditions, eg portable electrical equipment used outdoors, or in cramped spaces with a lot of earthed metal work, eg inside a boiler or bin.

The law

Look at the Electricity at Work Regulations 1989 for the full requirements.

Before you start

You may be able to do some jobs by using hand- or air-driven tools, but in practice electricity is so useful that no business could be without it.

Reduce the voltage

❍ lighting can run at 12 or 25 volts

❍ portable tools can run at 110 volts from an isolating transformer.

Provide a safety device

❍ a residual current device can act as a safety trip when there is a fault. This is not a substitute for a proper installation.

A safe installation

❍ provide enough socket outlets, if necessary, by using a multi-plug socket block - overloading sockets by using adapters can cause a fire

❍ fuses, circuit-breakers and other devices must be correctly rated for the circuit they protect

❍ there must be a switch or isolator near each fixed machine to cut off power in an emergency

❍ the mains switches must be readily accessible and clearly identified.

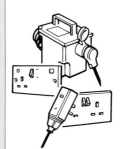
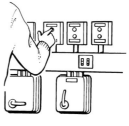

See the references section at the back of the book for details of publications which relate to **ELECTRICITY**

Insulation, protection and earthing

❍ power cables to machines must be insulated, eg sheathed and armoured or installed in conduit. Earth connections must be in good condition

❍ if you use a flexible cable you must always use a proper plug with the flex firmly clamped to stop the wires (particularly the earth) pulling out of the terminals

❍ some tools are double insulated for extra protection and these have only two wires (neutral and live). Make sure you connect them properly

❍ you must ensure plugs, sockets and fittings are sufficiently robust and adequately protected for the working environment

❍ replace frayed and damaged cables completely. Join lengths in good condition only by using proper connectors or cable couplers

❍ protect light bulbs, or other items which may be easily damaged in use

❍ you must use special protection where electrical equipment is used in flammable or dusty environments. Low voltage equipment (eg 12 volts) gives no protection against igniting flammable vapours. To choose the correct equipment you may need specialist advice

❍ when carrying or pouring organic powders (eg flour, tea dust) or flammable liquids, use closed metal containers and make sure all metal work is bonded and earthed

❍ for jobs like electrostatic paint spraying, make sure that both the work and anyone in the area are adequately earthed, eg by getting the operator and others to wear antistatic footwear, otherwise electrostatic charges can build up which can cause a spark.

Safe operation

You must not allow anyone to work on or near live equipment, unless it is unavoidable and special precautions are taken. Ask your inspector for advice.

Check that:

❍ suspect or faulty equipment is taken out of use, labelled 'Do not use' and kept secure until checked by a competent person

❍ tools and power sockets are switched off before plugging in or unplugging

❍ appliances are unplugged before cleaning or making adjustments.

Overhead electric lines

Contact with overhead electric lines accounts for half of the fatal electrical accidents each year. Electricity can flash over from overhead power lines even though plant and equipment may not touch them. Don't work under them where any equipment (eg ladders, a crane jib, a tipper lorry body, a scaffold pole) could come within 9 m of a power line without seeking advice. Consult your electricity company.

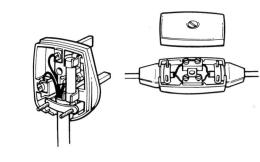

Maintenance

All electrical equipment, wiring installations, generators or battery sets and everything connected to them, must be maintained to prevent danger. This means carrying out checks and inspections and repairing and testing as necessary - how often will depend on the equipment you use and where you use it. (There are some essential references at the back of the book.) You may find it helpful to keep records of inspection and combined inspection and testing.

Don't forget hired or borrowed tools or equipment like floor polishers which may be used after the premises have closed.

❍ you must prevent access to electrical danger by keeping isolator and fuse box covers closed and (if possible) locked, with the key held by a responsible person

❍ anyone carrying out electrical work must be competent to do it safely. This may mean bringing in outside contractors; if so make sure they belong to a body which checks their work, such as the National Inspection Council for Electrical Installation Contracting (NICEIC)

❍ check that residual current circuit-breakers work by operating the test button regularly

❍ make sure that special maintenance requirements of waterproof or explosion-protected equipment have been written down and that someone is made responsible for carrying out the work without damaging the protection.

Please note: this is a model

Electric shock

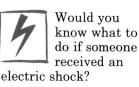 Would you know what to do if someone received an electric shock?

Knowing what to do should be part of your emergency procedures and first aid arrangements (see Chapter 16). Think about displaying a copy of the 'Electric shock placard', which shows you what to do.

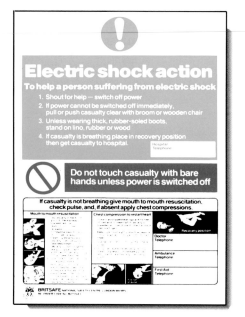

Underground cables

❍ consult your electricity company if you are likely to be digging near buried cables - they should know where these are

❍ always assume cables will be present when digging holes in the street, pavement or near buildings. If you have to work near services, use service plans, locators and safe digging practice to avoid danger.

SUBSTANCES

WHAT ARE THE HAZARDS?

Many substances can hurt you if they get into your body. Exposure can have an immediate effect and repeated exposure can damage your lungs, liver or other organs. Some substances may cause asthma and many can damage the skin.

Special care is needed when handling cancer-causing substances (carcinogens).

Chapter 14 covers flammable and explosive substances, while radiation hazards are dealt with in Chapter 11. Chapter 17 gives advice about watching out for possible symptoms of ill health.

The law

Look at the Control of Substances Hazardous to Health Regulations 1999 (COSHH) for the full requirements. Where you use something which might cause harm, this chapter will help you carry out the 'assessment' required by COSHH. It will help you decide if you need to make any changes.

It will also help you to decide if changes are not required, eg because the substance is a very low hazard or is used in trivial amounts, eg washing-up liquid or typing correction fluid. There are also special regulations covering asbestos and lead.

If you are a supplier, look at the Chemicals (Hazard Information and Packaging for Supply) Regulations 1994 (CHIP 2 as amended by CHIP 96, CHIP 97, CHIP 98 and CHIP 99), and the various 1996 regulations dealing with the carriage of dangerous goods by road and rail. The contents and hazards of the product must be indicated on the package or label. Safety data sheets must also be provided.

If you manufacture or import new chemicals, look at the Notification of New Substances Regulations 1993.

See the references section at the back of the book for details of publications which relate to **HARMFUL SUBSTANCES**

What are the risks?

You must carry out an assessment.

Consider:

❍ the hazards of substances or their ingredients - read the labels and safety data sheets. If in doubt contact your supplier

❍ the route into the body (breathed in, swallowed or taken in through the skin) and the worst result

❍ the concentration or conditions likely to cause ill health

❍ whether you know the first symptoms of over-exposure

❍ who could be exposed. Don't forget contractors and members of the public

❍ if they could be exposed accidentally, eg while cleaning, through spillage or if your controls fail

❍ how many people are involved

❍ how often they work with the substance

❍ how much they work with and how long for.

Substitute

Do you really need to use a particular substance? Can you use a safer material or change the process? If not, here is the best order of 'control measures'. You may need more than one.

Isolate or enclose

❍ put the harmful substances or process in a separate room or building or outside - but secure from the public

❍ reduce the amount used and number of people exposed, and the time they are exposed for

❍ use closed transfer and handling systems.

Use local exhaust ventilation

Use a local exhaust ventilation (LEV) system which sucks dust or vapour through a small hood or booth and takes it away from the worker.

A good system will:

❍ extract dust or vapour as close as possible to its source

❍ control contamination of the work area below the exposure limit for the material

❍ suck air away from the breathing zone of the operator - not through it

❍ have an adequate air flow at the source of the pollutant - 1 m/s at the face of a booth is a good guide

❍ have sufficient air flow inside ducts to prevent dust being deposited inside and blocking them

❍ have duct work with gently angled bends and junctions and tapered diameter changes

❍ make sure air is not vented back into the work area through roof lights or windows.

Special filters may be needed if the air is discharged outside or back into the room - ask your inspector for advice if in doubt.

General ventilation

A good supply and circulation of fresh air will help dilute minor contamination.

Good housekeeping

Simple precautions can cut exposure:

❍ do not store chemicals in open containers such as bottles or jam jars - make sure labels are not damaged, removed or covered up

❍ keep dangerous chemicals locked away

❍ clear up spillages quickly and safely

❍ have smooth work surfaces to allow easy cleaning

❍ clean regularly using a 'dust free' method such as a vacuum system with a high efficiency filter

❍ keep dusty materials, waste and dirty rags in covered containers

❍ do not let paste or drips dry out.

Exposure limits and air sampling

You must control the amount of dust or vapour in a worker's breathing zone to an acceptable level.

Dangerous dust and vapour is not always visible. Very small particles which you may not be able to see can get deep into the lungs and may be absorbed into the body, causing scars or ill health years later.

For many substances, limits have been set. They are listed in booklet EH40 *Occupational exposure limits*. This also explains how measurements are made.

In some cases, occupational hygienists may be needed to measure workers' exposure to toxic materials. They use air sampling techniques, but the way the process is carried out and the significance of the results need careful consideration - your inspector may be able to help.

Sampling may also be necessary to show conditions are safe, eg before allowing workers to enter tanks or vessels (see Chapter 6).

Good welfare and personal hygiene

Provide good washing and changing facilities. The Control of Lead at Work Regulations 1998 and the Control of Asbestos at Work Regulations 1987 have some particular requirements. See Chapter 2 for more general requirements.

Do not . . .

❍ smoke, eat or drink in chemical handling areas

❍ siphon or pipette hazardous chemicals by mouth - use a pump or hand-operated siphon

❍ transfer contamination, eg by putting pens and pencils in your mouth.

Do . . .

❍ remove protective clothing and wash hands before smoking, eating or drinking.

Personal protective clothing and equipment

 You must only use personal protective clothing and equipment as a last resort if you cannot control exposure in the ways outlined above. Chapter 18 gives further advice.

Lead

Work which exposes people to lead or its compounds is covered by the Control of Lead at Work Regulations 1998 and an Approved Code of Practice. Risks may arise when lead dust or fume is breathed in; powder, dust, paint or paste swallowed; or compounds taken in through the skin.

As well as obvious work such as high temperature melting, making batteries or repairing radiators, there may be risks from repair or demolition of structures which have been painted with lead-based paints.

You must assess the risk and where necessary provide control of the process, protective clothing, air sampling and health surveillance.

If you work with lead it is advisable to discuss the work with your inspectors and EMAS.

Maintaining the controls

Decide what needs to be done to ensure that the controls stay in place.

This will include:

❍ maintaining plant and equipment - all ventilation equipment must be examined and tested regularly by a competent person. Checks may include measuring the air speed or the pressures in the system, or air sampling in the work room. In general all LEV must be examined and tested every 14 months

❍ checking that people follow your rules

❍ checking that personal protective equipment is not worn out.

In special cases you must monitor exposure and carry out health surveillance (see Chapter 17).

Asbestos

Asbestos has been widely used,
eg as lagging on plant and pipework, in
insulation products such as fireproofing
panels, in asbestos cement roofing
materials, and as sprayed coating on
structural steelwork to insulate against fire
and noise.

All types of asbestos can be dangerous if
disturbed. The danger arises when asbestos
fibres as a very fine dust become airborne
and are breathed in. Exposure can cause
diseases such as lung cancer.

Well-sealed, undamaged asbestos is often
best left alone. Make sure that all asbestos
is sealed and protected against damage.
If you cannot seal and protect it and it is
likely to give off dust, you may need to
have it removed.

**If you have to work on asbestos you
must:**

❍ avoid breathing the dust - you may
need a respirator

❍ use the working methods and
precautions described in the
Asbestos Approved Codes of Practice, or
other equally safe methods.

**Asbestos has its own
regulations:**

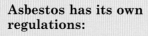

WARNING
CONTAINS
ASBESTOS

Breathing asbestos
dust is dangerous
to health

Follow safety
instructions

❍ all work with
asbestos and the
precautions
needed, including
respirators, are
covered by the
Control of Asbestos
at Work Regulations
1987

❍ the Asbestos (Licensing) Regulations 1983
prohibit contractors working on asbestos
insulation, asbestos board or asbestos

coating unless they have a licence issued by
HSE. This does not include asbestos cement
sheets

❍ the Asbestos (Prohibitions) Regulations
1992 prohibit the import and supply
(including second-hand articles) and use
of all types of asbestos except chrysotile
(white asbestos) - although they do prohibit
some uses of chrysotile.

Products containing asbestos carry a warning
label.

If you work with asbestos it is advisable to
discuss the work with your inspector and EMAS.

Information and training

You should tell workers:

❍ *the hazards*

❍ *how they could be affected*

❍ *what to do to keep themselves and others
safe, ie how the risks are to be controlled*

❍ *how to use control equipment and personal
protective equipment*

❍ *how to check and spot when things are wrong*

❍ *the results of any exposure monitoring or
health surveillance*

❍ *about emergency procedures
(see Chapter 16).*

Record and review

Except in very simple cases, you should keep a record of what you have found out and decided to do.

Write down:

❍ *which exposures need to be controlled*

❍ *how exposures are to be controlled*

❍ *how you will maintain control.*

Keep an eye on things. Changes in equipment, materials or methods may require you to review your earlier decisions. In any case you should take another look at your risks at least every five years.

Bacteria and viruses

Bacteria and viruses can:

❍ infect the body when they are breathed in, swallowed, or when they penetrate the skin

❍ cause allergic reactions.

Hazards include:

❍ legionnaires' disease - the bacteria causing this can be found in many recirculating water systems such as air-conditioning plant, cooling towers, industrial sprays and showers

❍ water-borne infections such as leptospirosis (Weil's disease). This is associated with rats, but can be spread by, eg cattle, and may affect, eg farmers, water industry workers

❍ infections through blood contact - a risk in, eg hairdressing, tattooing, health care

❍ diseases transmitted by living or dead animals, eg to farmers, pet shop workers

❍ diseases from people - a risk, eg for health care workers.

Most biological risks can be reduced by simple control methods. Sometimes immunisation may be needed. If you are in a risk group seek advice from EMAS.

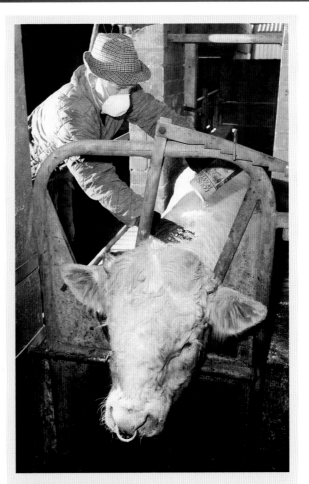

Skin problems

Dermatitis is a rash commonly affecting hands, forearms and legs. It may be caused by contact with chemicals, abrasives, cutting oils, solvents and resins.

Precautions include:

❍ reading the labels on containers

❍ keeping the workplace clean

❍ avoiding skin contact with the substance

❍ wearing impermeable gloves - don't let the substance get inside them

❍ keeping skin clean - dry thoroughly after washing and don't use abrasives or solvents

❍ not letting glues or resins harden on the skin and using after-care cream

❍ getting first-aid treatment for minor cuts and keeping them covered

❍ checking skin regularly. Seek medical advice at once for any rashes.

SUBSTANCES

WHAT ARE THE HAZARDS?

Some gases, liquids and solids can cause explosions or fire. For a fire to start, fuel, air and a source of ignition are needed.

Common materials may burn violently at high temperature in oxygen-rich conditions, eg when a gas cylinder is leaking.

Some dusts form a cloud which will explode when ignited. A small explosion can disturb dust and create a second explosion severe enough to destroy a building.

Serious explosions can occur in plant such as ovens, stoves and boilers (see Chapter 5).

Some materials are explosives and need special precautions and licensing arrangements.

Some flammable liquids and substances are also corrosive or toxic and may pose risks to health (see Chapter 13).

The law

There is little general law and most of this section is good advice, but for flammable liquids and LPG look at the Highly Flammable Liquids and Liquefied Petroleum Gases Regulations 1972.

If you store even small quantities (eg 15 litres) of petrol or petroleum mixtures you may need a licence under the Petroleum Acts - ask your local authority.

If you are a supplier look at the Chemicals (Hazard Information and Packaging for Supply) Regulations 1994 (CHIP 2 as amended by CHIP 96, CHIP 97, CHIP 98 and CHIP 99), and the the various 1996 regulations dealing with the carriage of dangerous goods by road and rail.

If you manufacture or import new chemicals, look at the Notification of New Substances Regulations 1993.

See the references section at the back of the book for details of publications which relate to **FLAMMABLE AND EXPLOSIVE SUBSTANCES**

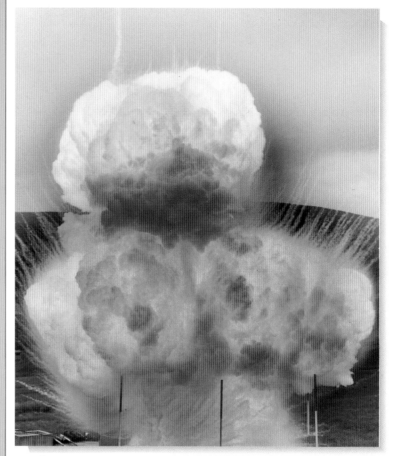

Before you start

The supplier's safety data sheet will help you decide how to handle these substances.

Think about:

❑ doing the job another way

❑ using liquids with higher flash points. Look at the data sheets - remember a high flash point is safer than a low one

❑ reducing the amounts you keep on site

❑ checking with the supplier about any special precautions which may be needed when certain materials are delivered in bulk

❑ checking container labels and consignment notes to make sure that goods are supplied as ordered.

Explosives

Specific control is applied to most explosives, including fireworks and safety cartridges. You may need a licence or other permit - contact your inspector for advice.

Some substances like organic peroxide and other oxidisers can explode if they are not stored and handled properly. Check labels and safety data sheets.

Storage

Some chemicals react dangerously together. Such classes of material should be stored correctly, eg oxidising substances should be kept apart from flammable ones.

Use the information from the supplier and the package label to decide storage arrangements. Materials can be separated by distance, by a physical barrier or (sometimes) by other non-reactive materials.

Good storage will:

❍ be separate from process areas (where fire or leakage is more likely to occur)

❍ be in a safe, well-ventilated place, isolated from buildings

❍ prevent incompatible chemicals being mixed, eg by spillage, damage to packaging or by wetting during fire fighting

❍ reduce the risk of damage, eg by lift truck, and by ensuring that cylinders are secured and stored upright

❍ prevent rapid spread of fire or smoke, or liquid or molten substances, eg by the store being made of fire-resisting material

❍ exclude sources of ignition, eg static electricity, unprotected electrical equipment, cigarettes and naked flames

❍ include empty drums or cylinders as well as full ones - the risk can be just as great.

Housekeeping

❍ remove grease frequently from ducts, such as kitchen ventilators and cooker extractor hoods

❍ keep the workplace tidy and free from old containers etc. Plastic foam crumb and off-cuts are a particular hazard

❍ contaminated clothing or containers need careful disposal

❍ keep flammable waste secure from vandals.

Flammable liquids

The safest place to store any flammable liquids and substances is in a separate building or in a safe place in the open air. If highly flammable liquids have to be stored inside workrooms you must not keep more than 50 litres and they should be kept on their own in a special metal cupboard or bin. Larger stocks should be held in a fire-resisting store with spillage retention and good ventilation.

If you run a factory you must:

❍ minimise the amount kept at the workplace

❍ dispense and use in a safe place with adequate natural or mechanical ventilation

❍ keep containers closed, eg use safety containers with self-closing lids and caps

❍ contain spillages, eg by dispensing over a tray and having absorbent material handy

❍ control ignition sources, eg naked flames and sparks, and make sure that 'no smoking' rules are obeyed, especially when spraying highly flammable liquids

❍ keep contaminated material in a lidded metal bin which is emptied regularly

❍ get rid of waste safely, eg burn rubbish in a suitable container well away from buildings. Have fire extinguishers on hand. Don't burn aerosol cans and don't 'brighten' fires with flammable liquids.

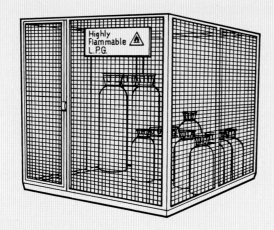

Gas cylinders

Storage and use:

❍ store both full and empty cylinders in a secure outside compound where possible

❍ store with valves uppermost, particularly where they contain liquid like acetylene

❍ don't store them below ground level or near to drains or basements - most gases are heavier than air

❍ protect cylinders from damage, eg by chaining unstable cylinders in racks or on special trolleys

❍ use the right hoses, clamps, couplers and regulators for the particular gas and appliance

❍ turn off cylinder valves at the end of each day's work

❍ change cylinders away from sources of ignition, in a well-ventilated place

❍ avoid welding flame 'flash-back' into the hoses or cylinders by training operators in correct lighting up and working procedures and by fitting non-return valves and flame arresters

❍ use soap or detergent/water solution, never a flame, to test for leaks

❍ before welding and similar work, remove or protect flammable material

❍ where possible, position gas cylinders on the outside of buildings and pipe through to appliances or processes

❍ make sure that rooms where appliances, eg LPG heaters, are used have sufficient ventilation high up and low down, which is never blocked up to prevent draughts.

Dust explosions

Do you have a dusty process? Is the dust flammable? Examples include aluminium powder, flour, bone-meal, cotton fly, paper dust, polystyrene and fine sawdust.

If so you must:

❍ keep plant dust-tight and frequently checked and cleaned

❍ avoid buildup of dust, eg reduce the number of ledges and horizontal surfaces on which dust may settle and use exhaust ventilation with suitable dust collectors as necessary

❍ control sources of heat such as welding, space heaters and smoking

❍ reduce sparking by using dust-tight electrical equipment, by earthing sources of static electricity and by using magnets to catch any stray pieces of metal before they get into the process

❍ take explosion protection measures, by providing explosion vents or a plant structure strong enough to withstand an explosion

❍ make sure explosion vents discharge safely.

You can reduce the effects of an explosion by using lightweight construction for buildings which house dangerous plant.

Flammable solids

Plastic foams are high fire risk and need careful control, both for storage and in the workroom - treat them like other flammable materials.

❍ use non-sparking tools when scraping deposits from spray booths etc.

Special cases

You must notify HSE:

❍ if you use or store certain listed 'major hazard' dangerous substances. You will have to demonstrate to your inspector that you are operating safely

❍ if you market more than one tonne per year of a new substance.

The discharge of effluent, disposal and transport of waste, and the emission of smoke and chemicals to the atmosphere may need special precautions or authorisation. Contact your inspector or local authority for advice.

Oxygen

Common materials may burn violently at high temperature in the presence of oxygen.

❍ never use oxygen to 'sweeten' the atmosphere

❍ make sure there are no leaks, especially in confined areas, and don't use oxygen to operate compressed air equipment

❍ keep oxygen cylinders free from grease and other combustible materials and don't store them with flammable gases or materials.

If you are a supplier

You must:

❍ provide safety data sheets and other information for users

❍ if supplying a new substance see the 'Special cases' section opposite

❍ arrange for any necessary testing and research so that substances can be used safely at work

❍ choose packaging which provides protection for users and during conveyance and transport

❍ provide labels which give adequate information about the risks and necessary precautions

❍ remember you need to look at the Chemicals (Hazard Information and Packaging for Supply) Regulations 1994 (CHIP 2 as amended by CHIP 96, CHIP 97, CHIP 98 and CHIP 99), and the various 1996 regulations dealing with the carriage of dangerous goods by road and rail.

Transporting materials

If you transport and deliver materials off site, then you must:

❍ ensure the packages are suitable and correctly labelled as required for carriage by road, rail, air or sea transport

❍ ensure that the vehicle is suitable for the purpose

❍ provide appropriate fire-fighting equipment

❍ fit hazard panels/plates on delivery vehicles as required

❍ check compatibility of loads

❍ provide written information for the driver

❍ train vehicle drivers on their duties, the hazards and risks involved and the necessary emergency procedures as required by the Carriage of Dangerous Goods by Road (Driver Training) Regulations 1996.

Emergencies

Consider what could go wrong:

❍ could staff accidentally mix incompatible chemicals, eg bleach with other cleaners?

❍ are you prepared for a large leak or spillage?

❍ what about hazardous by-products? Could mixing of waste chemicals in the drains cause a hazardous reaction or pollution?

❍ are any special first-aid facilities or equipment required?

❍ could emergency water supplies freeze up in winter?

More information is given on accidents and emergencies in Chapter 16. General fire precautions are covered in Chapter 2.

PROCEDURES

WHAT ARE THE HAZARDS?

The previous chapters have looked at particular hazards. You now need to think more generally about the safe procedures and systems of work needed to deal with them.

The law

Section 2 of the HSW Act requires 'safe systems of work', but does not go into detail. This chapter gives advice on what this phrase means in practice.

Look at the Management of Health and Safety at Work Regulations 1992 for the requirement to make appropriate arrangements for managing health and safety.

You may need to consider the Confined Spaces Regulations 1997.

See the references section at the back of the book for details of publications which relate to **SAFE SYSTEMS**

Clear procedures

Having clear procedures helps you get things right and check that work is being done safely. For serious hazards and risks it is worth writing them down, eg a written 'permit to work'. This may not be necessary for many ordinary jobs.

When looking at your systems, don't forget:

- ❍ routine work (including setting up and preparation, finishing off and cleaning activities)

- ❍ less routine work, eg maintenance

- ❍ emergencies, eg fire, spillages or plant breakdown (emergency procedures are covered in Chapter 16).

Think of all the different things you do. Ask those who do the job to tell you what they actually do and how they do it. Get them to help identify the hazards and risks.

Safe procedures

Think about:

- ❍ who is in charge of the job?

- ❍ do the responsibilities overlap with anyone else's?

- ❍ is there anything which is not someone's responsibility?

- ❍ has anyone checked that the equipment, tools or machines are right for the job?

- ❍ are safe ways of doing the job already in place?

- ❍ could this job interfere with the health and safety of others?

- ❍ are safe working procedures laid down for the job. Is there any guidance which may help you?

- ❍ have people been trained and instructed in the use and limitations of equipment?

- ❍ if the job cannot be finished today can it be left in a safe state?

- ❍ are clear instructions available for the next shift?

- ❍ are the production people aware of what maintenance staff are doing and vice versa?

- ❍ what might go wrong, eg accident, explosion, food poisoning, electrocution, fire, release of radioactivity, chemical spill?

Permits to work

Simple instructions or lock-off procedures are adequate for most jobs, but some require extra care. A 'permit to work' states exactly what work is to be done and when, and which parts are safe. A responsible person should assess the work and check safety at each stage. The people doing the job sign the permit to show that they understand the risk and precautions necessary.

Examples of high risk jobs where a written 'permit to work' procedure may need to be used include:

❍ entry into vessels, confined spaces or machines

❍ hot work which may cause explosion or fire

❍ construction work or the use of contractors

❍ cutting into pipework carrying hazardous substances

❍ mechanical or electrical work requiring isolation of the power source, eg before work inside large machines, if locking off is not good enough

❍ work on plant, mixers, boilers etc which must be effectively cut off from the possible entry of fumes, gas, liquids or steam

❍ testing for dangerous fumes or lack of oxygen before entering an unventilated pit or silo

❍ vacuuming the inside of an empty grain silo to remove dust which might explode, before hot cutting a hole in the side.

When the risk is high your precautions should be 100% correct. If in doubt discuss them with your inspector.

Lock-off procedures

Before working on plant or equipment, isolate machines from the main power supply by locking off the power. Usually this is done by using a separate electrical switch.

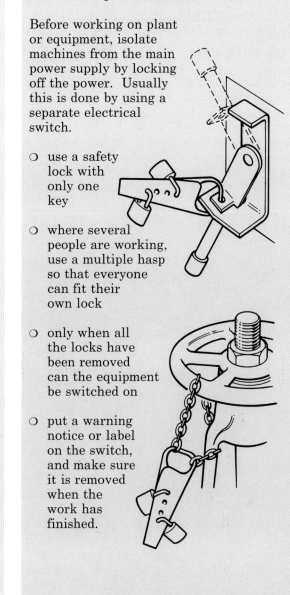

❍ use a safety lock with only one key

❍ where several people are working, use a multiple hasp so that everyone can fit their own lock

❍ only when all the locks have been removed can the equipment be switched on

❍ put a warning notice or label on the switch, and make sure it is removed when the work has finished.

Check your systems

You cannot rely on your systems always being right. Check that your rules and procedures not only deal with all the risks but are also being followed - particularly if people are working outside 'normal hours' with less supervision than usual.

❍ are all risks covered?

❍ do people follow procedures and do they have any ideas for improvements? Do you ask them?

❍ are there any training gaps?

❍ is the level of supervision right?

PROCEDURES

BEFORE THE EVENT

In any business, things sometimes go wrong. You need to be ready to deal with these unplanned events. Look at incidents which have caused injuries and ill health or other damage - what can you learn? Think about emergencies - plan for the worst that can happen. You must have the right first-aid arrangements. Some events need to be reported - this chapter tells you how to do this.

Fire precautions are covered in Chapter 2.

The law

The Management of Health and Safety at Work Regulations 1992 cover emergencies.
The requirements for first aid are in the Health and Safety (First Aid) Regulations 1981, and for reporting incidents in the Reporting of Injuries, Diseases and Dangerous Occurrences Regulations 1995 (RIDDOR).

The Dangerous Substances (Notification and Marking of Sites) Regulations 1990 cover sites where at least 25 tonnes of dangerous substances are held.

This chapter outlines the main legal points and gives some extra advice.

See the references section at the back of the book for details of publications which relate to **ACCIDENTS AND EMERGENCIES**

Emergency procedures

When things go wrong, people may be exposed to serious and immediate danger. Special procedures are necessary for emergencies such as serious injuries, explosion, flood, poisoning, electrocution, fire, release of radioactivity and chemical spills.

Write an emergency plan if a major incident at your workplace could involve risks to the public, rescuing employees or co-ordination of emergency services.

Think about:

❍ the worst that can happen if things go wrong?

❍ how the person in charge and others will deal with the problems? You should look at any particular responsibilities and training needs

❍ if everyone is adequately prepared - could emergency services get to the site?

Points to include in emergency procedures:

❍ consider what might happen and how the alarm will be raised. Don't forget night and shift working, week-ends and (possibly) times when the premises are closed, eg holidays

❍ plan what to do, including how to call the emergency services. Assist the emergency services by clearly marking your premises from the road. Consider drawing up a simple plan marked with the location of hazardous items

❍ if you have at least 25 tonnes of dangerous substances you must notify the fire authority and put up warning signs

❍ decide where to go to reach a place of safety or to get rescue equipment. Provide emergency lighting if necessary

❍ you must make sure there are enough emergency exits for everyone to escape quickly, and keep emergency doors and escape routes unobstructed and clearly marked

❍ *nominate competent persons to take control*

❍ decide who are the other key people such as first aiders?

❍ plan essential actions such as emergency plant shut-down or making processes safe. Clearly label important items like shut-off valves and electrical isolators etc

❍ you must train everyone in emergency procedures. Don't forget the needs of people with disabilities.

Reporting injuries and other events

RIDDOR applies to all employers and the self-employed and covers everyone at work.

The main points are that you must:

❍ notify your inspector immediately, normally by telephone, if anybody dies, receives a major injury or is seriously affected by, eg an electric shock or poisoning

❍ notify your inspector immediately if there is a dangerous occurrence, eg a fire or explosion which stops work for more than 24 hours, or an overturned crane

❍ confirm in writing within 10 days on form F2508

❍ report within 10 days (on form F2508) injuries which keep an employee off work or unable to do their normal job for more than three days

❍ report certain diseases suffered by workers who do specified types of work as soon as possible on learning about the illness. Use form F2508A

❍ if you supply, fill or import flammable gas in reusable containers, you must notify HSE immediately of any death or injury connected in any way with the gas supplied and confirm the notification with a report on F2508 within 14 days

❍ keep details of the incident.

Investigating events

When an accident happens:

❍ take any action required to deal with the immediate risks, eg first aid, put out the fire, isolate any danger, fence off the area

❍ assess the amount and kind of investigation needed - if you have to disturb the site, take photographs and measurements first

❍ investigate - find out what happened and why

❍ take steps to stop something similar happening again

❍ also look at near misses and property damage. Often it is only by chance that someone wasn't injured.

Learn from the experience - what changes do you need to make?

First aid

You must have:

❍ someone who can take charge in an emergency. An appointed person must be available whenever people are at work

 ❍ a first-aid box

❍ notices telling people where the first-aid box is and who the appointed person is

❍ a trained first aider and a first-aid room if your work gives rise to special hazards, eg using a particularly toxic material.

As your company grows, look again at your need for qualified first aiders. They must have the right training and are given a certificate valid for three years - after that a refresher course and re-examination is necessary. Training organisations are registered with EMAS - ask your local Employment Medical Adviser.

Checklist

To help with your investigation, find out:

 ❍ details of injured personnel

❍ details of injury, damage or loss

❍ what was the worst that could have happened? Could it happen again?

❍ what happened? Where? When? What was the direct cause?

❍ were there standards in place for the premises, plant, substances, procedures involved?

❍ were they adequate? Were they followed?

❍ were the people up to the job? Were they competent, trained and instructed?

❍ what was the underlying cause? Was there more than one?

❍ what was meant to happen and what were the plans? How were the people organised?

❍ if inspection would have picked up the problem earlier?

❍ if it had happened before? If so, why weren't the lessons learnt?

Most accidents have more than one cause so don't be too quick to blame individuals - try to deal with the root causes.

PEOPLE

WHAT ARE THE HAZARDS?

More people die from work-related diseases than from workplace accidents. This chapter deals with health risks in general, mental health and stress, drugs and alcohol, and smoking.

See Chapter 11 for radiation risks and Chapter 13 for information on health risks from hazardous substances. Emergency arrangements and first aid are covered in Chapter 16.

The law

Health surveillance is covered by the Management of Health and Safety at Work Regulations 1992, and COSHH in relation to substances. It is only needed in special cases. The rest of this chapter is advice.

See the references section at the back of the book for details of publications which relate to **HEALTH CARE**

Check for ill health

To help prevent work-related diseases:

❍ check whether there are any known health risks from the work you do

❍ listen to complaints about ill health - particularly if there are a number of them, or a lot of sickness absence in the same area, or as a result of the same activity

❍ encourage people to tell their doctor about the work they do if they think their work might be affecting them, eg the type of work, materials used, exposure to dirty water or sewage

❍ if in doubt get further advice. Contact EMAS at your local HSE office.

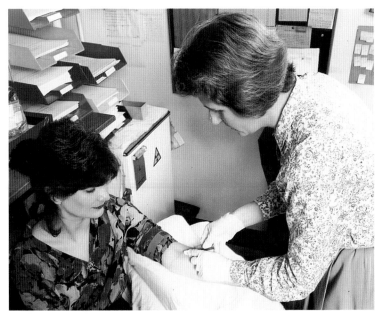

Health surveillance

Health surveillance means having a system to look for early signs of ill health caused by substances and other hazards at work. It includes keeping health records for individuals and may include medical examinations and testing of blood or urine samples, so that corrective action can be taken.

Health surveillance is not required for most workers. You must decide whether it is needed for what you do. If in doubt, eg if there are known health risks from the work, get advice.

Medical examinations or health surveillance are required by law for some jobs, eg commercial diving, work with asbestos insulation and work with some chemicals.

Special conditions apply to some people who are particularly at risk, such as pregnant women whose job may expose them to lead or ionising radiation.

You must tell employees about any symptoms they should look out for.

Mental health and stress

Many people suffer some form of mental health problem at some time and it is a significant reason for absence from work. Most problems do not last and most people do not have to give up work.

❍ work can have a beneficial effect on mental health, but can also cause worries

❍ too little personal control over the work, not being allowed to use all your skills fully, being overworked or underworked, and boring work can all contribute to stress

❍ learn to recognise signs of stress and encourage employees to discuss problems openly - are people distracted, tense or worried?

❍ give help. This may be simple sympathetic reassurance or practical advice, or in some cases counselling or psychiatric treatment. Don't delay - this can make matters much worse.

Drugs and alcohol

Abuse of alcohol, drugs and other substances can affect work performance and safety.

❍ learn to recognise the signs and encourage workers to seek help

❍ be supportive

❍ if you decide strict standards are needed because of safety-critical jobs, then agree procedures with workers in advance. Disciplinary action may be needed where safety is critical.

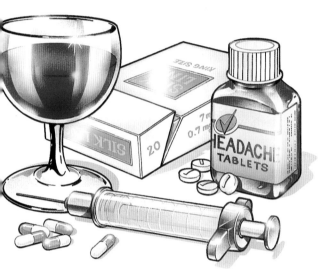

Passive smoking

This is breathing in other people's tobacco smoke. It can damage health, make asthma worse and cause lung cancer.

Think about:

❍ encouraging and helping smokers to give up

❍ agreeing some rules with the workforce to protect non-smokers

❍ if possible providing a smoke-free workplace - at least consider separate 'break' areas. Any rest rooms or rest areas you provide must be arranged to avoid discomfort to non-smokers.

PEOPLE

WHAT ARE THE HAZARDS?

Even where engineering controls and safe systems of work have been applied, some hazards might remain. These include injuries to the lungs, eg from breathing in contaminated air; the head and feet, eg from falling materials; the eyes, eg from flying particles or splashes of corrosive liquids; the skin, eg from contact with corrosive materials; the body, eg from extremes of heat or cold.

Personal protective equipment (PPE) is needed in these cases to reduce the risk.

This chapter tells you the main things you must do and gives some information about different kinds of protection.

The law

The Personal Protective Equipment at Work Regulations 1992 give the main requirements, but other special regulations cover lead, asbestos, hazardous substances (see Chapter 13), noise (see Chapter 9) and radiation (see Chapter 11).

See also the Construction (Head Protection) Regulations 1989.

See the references section at the back of the book for details of publications which relate to

PERSONAL PROTECTIVE EQUIPMENT

The last resort

Use personal protective equipment (PPE) only as a last resort; wherever possible engineering controls and safe systems of work should be used instead. If PPE is still needed it must be provided free by the employer.

Selection and use

You must consider:

❍ who is exposed and to what?

❍ for how long?

❍ to how much?

You must:

❍ choose good quality products made to a recognised standard - suppliers can advise

❍ choose equipment which suits the wearer - consider size, fit and weight. If you let the users help choose it, they will then be more likely to use it

❍ make sure it fits properly - note in particular the problem of creating a good seal if a respirator user has a beard

❍ make sure that if more than one item of PPE is being worn they can be used together, eg a respirator may not give proper protection if air leaks in around the seal because the user is wearing safety glasses

❍ instruct and train people in its use. Tell them why it is needed, when to use it and what its limitations are.

Remember that PPE is a last resort but must be worn when needed.

Other advice:

❍ many solvents quickly go through rubber-based materials. Few materials protect you if soaked in hazardous chemicals

❍ never allow exemptions for those jobs which take 'just a few minutes'

❍ check with your supplier - explain the job to them

❍ if in doubt seek further advice from your inspector or other specialist adviser.

Maintenance

Equipment must be properly looked after and stored when not in use, eg in a dry, clean cupboard. It must be cleaned and kept in good repair.

Think about:

❍ using the right replacement parts which match the original, eg respirator filters

❍ keeping replacement PPE available

❍ who is responsible for maintenance and how it is to be done

❍ having a supply of disposable suits which are useful for dirty jobs where laundry costs are high, eg for visitors who need protective clothing.

Employees must make proper use of PPE and report its loss or destruction or any fault in it.

Finally

❍ check regularly that PPE is used or find out why not. Safety signs can be a useful reminder

❍ make a note of any changes in equipment, materials and methods - you may need to update what you provide.

Eyes

Hazards: chemical or metal splash, dust, projectiles, gas and vapour, radiation.

Options: spectacles, goggles, face screens, helmets.

Note: make sure the eye protection chosen has the right combination of impact/dust/splash/molten metal etc protection for the task.

Head and neck

Hazards: impact from falling or flying objects, risk of head bumping, hair entanglement, chemical drips or splash, climate or temperature.

Options: helmets, bump caps, hairnets, sou'westers and cape hoods.

Note: some safety helmets incorporate or can be fitted with specially-designed breathing or hearing protection. Don't forget neck protection, eg scarves for use during welding.

Ears

Hazards: impact noise, high intensities (even if short exposure), pitch (high and low frequency).

Options: ear-plugs or muffs.

Note: ear-plugs may be pre-shaped or individually moulded in rubber or plastic, or disposable and made of compressible plastic foam, glass-down etc. Take advice to make sure they can reduce noise to an acceptable level. Fit only specially-designed ear-muffs over safety helmets. See Chapter 9 for more information on noise.

Hands and arms

Hazards: abrasion; temperature extremes; cuts and punctures; impact; chemicals; electric shock; skin irritation, disease or contamination; vibration; risk of product contamination.

Options: gloves, gauntlets, mitts, wrist cuffs, armlets.

Note: don't wear gloves or mitts when operating machines such as bench drills where the gloves might get caught. Some materials are quickly penetrated by chemicals - care in selection is needed. Use skin conditioning cream after work with water or fat solvents. Barrier creams provide limited protection. Disposable or cotton inner gloves can reduce sweating.

Feet and legs

Hazards: wet; electrostatic build-up; slipping; cuts and punctures; falling objects; heavy loads; metal and chemical splash; vehicles.

Options: safety boots and shoes with steel toe caps (and steel mid-sole), gaiters, leggings, spats and clogs.

Note: footwear can have a variety of sole patterns and materials to prevent slips in different conditions; can have oil or chemical-resistant soles; and can be anti-static, electrically conductive or insulating. There is a variety of styles including 'trainers' and ankle supports. Avoid high-heeled shoes and open sandals.

Lungs

Hazards: dusts, gases and vapours.

Options: disposable respirators; half masks or full face mask respirators fitted with a filtering cartridge or canister; powered respirators blowing filtered air to a mask, visor, helmet, hood or blouse; fresh air hose equipment; breathing apparatus (self-contained and fresh air line types).

Note: The right type of respirator filter must be used as each is effective for only a limited range of substances. Cartridges and canisters have only a limited life. Where there is a shortage of oxygen or any danger of losing consciousness due to exposure to high levels of harmful fumes, use only breathing apparatus - never use a filtering cartridge. All equipment should be suitable for its purpose and meet the necessary standards.

Whole body

Hazards: heat; cold; bad weather; chemical or metal splash; spray from pressure leaks or spray guns; impact or penetration; contaminated dust; excessive wear or entanglement of own clothing.

Options: conventional or disposable overalls, boiler suits, warehouse coats, donkey jackets, aprons, chemical suits, thermal clothing.

Note: the choice of materials includes non-flammable; anti-static; chain-mail; chemically impermeable; and high visibility. Don't forget other protection, like safety harnesses or life-jackets.

Emergency equipment

Careful selection, maintenance and regular operator training is needed for equipment like compressed air escape breathing apparatus, artificial respirators and safety ropes or harnesses.

Here is an example of a chain-saw user who has:

○ a safety helmet - replace at intervals recommended by the manufacturer, eg every 2 to 3 years

○ ear defenders

○ eye protection

○ clothing - should be close fitting

○ gloves - with protective pad on the back of the left hand

○ protection for legs - incorporating loosely-woven long nylon fibres or similar material. All round leg protection is recommended

○ chain-saw operator boots - the casual user may obtain adequate protection by a combination of protective spats and industrial steel toe-capped safety boots.

PEOPLE

WHAT ARE THE HAZARDS?

People are a danger to themselves or others if they cannot do their jobs correctly. This can be because they are in jobs for which they are unsuited and/or are not competent.

In these cases, the hazards described in earlier chapters may not be controlled and an accident may happen.

The law

The Management of Health and Safety at Work Regulations 1992 set down some key general requirements.

Some groups, such as young people, new and expectant mothers, and those with a disability, need special consideration.

Children under school leaving age are generally not allowed to work in industrial activities and there are also restrictions on the part-time employment of children in other jobs.

Women and young persons (aged 16 to 18 years) are also prohibited from doing certain types of high risk work, eg melting lead scrap.

If in doubt seek advice from your inspector.

See the references section at the back of the book for details of publications which relate to **SELECTION AND TRAINING**

Selection

❍ identify jobs which place particular physical or mental demands on people - can these be changed to cut out or reduce the demands?

❍ where you cannot do this you need to select people to meet the demands

❍ identify the essential health requirements of a job and use them during recruitment, eg do not employ someone with epilepsy to climb ladders, although they may be able to do other work

❍ for some jobs, eg driving heavy goods vehicles, the law requires medical examinations, but pre-employment medical checks are not legally required for most jobs

❍ people returning after illness may need help readjusting to their jobs - seek specialist medical advice if necessary.

Training

 You must ensure all employees have health and safety training. This has to be repeated periodically or when changes are made.

Also think about:

❍ the needs of managers and supervisors

❍ any legal requirements for specific job training, eg lift trucks, first aid

❍ the special needs of young people, new recruits, trainees and part-time employees

❍ existing workers and those moving jobs

❍ training needs identified during investigations

❍ are there any standards of competence for what you do? eg NVQs or SVQs accredited by the National Council for Vocational Qualification or SCOTVEC, the Scottish Vocational Education Council

❍ can you give the training yourself or do you need some outside help?

REFERENCES

FURTHER READING

The following list of HSE publications is only a small selection of those available - a comprehensive list is available from HSE Books.

Free leaflets and other free items which are available from HSE Books are marked with an asterisk*.

For some leaflets, single copies are free, but multiple orders are priced.

While every effort has been made to ensure the accuracy of the references listed in this publication, their future availability cannot be guaranteed.

Abbreviations

BS	British Standard
COP	Code of Practice
CS	Guidance Note: Chemical safety
EH	Guidance Note: Environmental health
GS	Guidance Note: General series
MS	Guidance Note: Medical series
PM	Guidance Note: Plant and machinery
HSC	Health and Safety Commission leaflet
HSE	Health and Safety Executive leaflet
HSG	Health and Safety (Guidance) booklet
HSR	Health and Safety (Regulations) booklet
IAC	Industry Advisory Committee leaflet/booklet
INDG	Industry Advisory (General) leaflet
INDS	Industry Advisory (Special) leaflet
L	Legal series

ORGANISING FOR SAFETY

1 Managing health and safety

HSG65
Successful health and safety management
ISBN 0 7176 1276 7

HSG96
The costs of accidents at work
ISBN 0 7176 1343 7

HSG183
Five steps to risk assessment
ISBN 0 7176 1580 4

L1
A guide to the HSW Act
ISBN 0 7176 0441 1

L21
Management of Health and Safety at Work Regulations 1992.
Approved Code of Practice
ISBN 0 7176 0412 8
(Leaflet MISC079 is a free update of the above Regulations)

L87
Safety representatives and safety committees
ISBN 0 7176 1220 1

Poster
Health and safety law: what you should know
ISBN 0 7176 1380 1

Managing contractors: A guide for employers
ISBN 0 7176 1196 5

Managing health and safety: An open learning workbook for managers and trainers
ISBN 0 7176 1153 1

Writing your health and safety policy statement: Guide to preparing a safety policy statement for a small business
ISBN 0 7176 0424 1

You can do it. The what, why and how of improving health and safety - a self help guide
ISBN 0 7176 0726 7

* HSC6
Writing a safety policy statement: Advice to employers

* HSC13
Health and safety regulation - a short guide

* HSC14
What to expect when a health and safety inspector calls

* HSE4
Employers' Liability (Compulsory Insurance) Act 1969: a guide for employers

* HSE34
HSE and you

* HSE35
HSE - working with employers

* HSE36
Employers Liability (Compulsory Insurance) Regulations: a guide for employees and their representatives

* INDG163(rev1)
Five steps to risk assessment

* INDG208
Be safe - save money. The costs of accidents - a guide for small firms

* INDG218
A guide to risk assessment requirements: Common provisions in health and safety law

* INDG226
Homeworking: Guidance for employers and employees on health and safety

* INDG244
Workplace health, safety and welfare: A short guide for managers

* INDG259
An introduction to health and safety

* INDG275
Management of health and safety: five steps to success

PREMISES

2 The workplace

HSG38
Lighting at work
ISBN 0 7176 1232 5

HSG55
Health and safety in
kitchens and food
preparation areas
ISBN 0 11 885427 5

HSG57
Seating at work
ISBN 0 7176 1231 7

HSG67
Health and safety in motor
vehicle repair
ISBN 0 7176 0483 7

HSG70
Control of legionellosis
including legionnaires'
disease
ISBN 0 7176 0451 9

HSG76
Health and safety in retail
and wholesale warehouses
ISBN 0 11 885731 2

HSG90
VDUs: an easy
guide to the
Regulations
ISBN 0 7176 0735 6

HSG104
Health and safety in
residential care homes
ISBN 0 7176 0673 2

HSG122
New and expectant
mothers at work: A guide
for employers
ISBN 0 7176 0826 3

HSG129
Health and safety in
engineering workshops
ISBN 0 7176 0880 8

HSG132
How to deal with sick
building syndrome
ISBN 0 7176 0861 1

HSG155
Slips and trips: Guidance
for employers on
identifying hazards and
controlling risks
ISBN 0 7176 1145 0

HSG179
Managing health and
safety in swimming pools
ISBN 0 7176 1388 7

HSG192
Charity and voluntary
workers: a guide to health
and safety at work
ISBN 0 7176 2424 2

HSG194
Thermal comfort in the
workplace: guidance for
employers
ISBN 0 7176 2468 4

L24
Workplace health, safety
and welfare. Workplace
(Health, Safety and
Welfare) Regulations 1992.
Approved Code of Practice
and guidance
ISBN 0 7176 0413 6

L26
Display screen equipment
work. Health and Safety
(Display Screen Equipment)
Regulations 1992.
Guidance on Regulations
ISBN 0 7176 0410 1

L64
Safety signs and signals:
Guidance on Regulations.
Health and Safety (Safety
Signs and Signals)

Regulations 1996
ISBN 0 7176 0870 0

Fire safety:
an employer's guide
ISBN 0 11 341229 0

* INDG36 (rev1)
Working with VDUs

* INDG90(rev)
If the task fits:
ergonomics at work

* INDG173
Officewise

* INDG184
Signpost to the Health
and Safety (Safety Signs
and Signals) Regulations
1996

* INDG225
Preventing slips, trips and
falls at work

* INDG244
Workplace health, safety
and welfare: a short guide

* INDG293
Welfare at work: guidance
for employers on welfare
provisions

3 Building work

HSG33
Health and safety in
roof work
ISBN 0 7176 1425 5

HSG47
Avoiding danger from
underground services
ISBN 0 7176 0435 7

HSG144
Safe use of vehicles on
construction sites
ISBN 0 7176 1610 X

HSG150
Health and
safety in
construction
ISBN 0 7176 1143 4

HSG151
Protecting the public: Your
next move
ISBN 0 7176 1148 5

HSG168
Fire safety in construction:
guidance for clients,
designers and those
managing and carrying out
construction work
involving significant fire
risks
ISBN 0 7176 1332 1

HSG185
Health and safety in
excavations
ISBN 0 7176 1563 4

L24
Workplace health, safety
and welfare. Workplace
(Health, Safety and
Welfare) Regulations 1992.
Approved Code of Practice
and guidance
ISBN 0 7176 0413 6

L54
Managing construction for
health and safety (CDM
Regulations). Approved Code
of Practice
ISBN 0 7176 0792 5

PM63
Inclined hoists used in
building and construction
work
ISBN 0 11 883945 4

A guide to managing health
and safety in construction
ISBN 0 7176 0755 0

Designing for health and
safety in construction
ISBN 0 7176 0807 7

* INDG220
A guide to the Construction
(Health, Safety and Welfare)
Regulations 1996

* INDG284
Working on roofs

PLANT AND MACHINERY

4 Machinery safety

BS 5304:1988
Safety of machinery

BS EN 292:1991
Safety of machinery.
Basic concepts, general
principles for design

HSG17
Safety in the use of abrasive
wheels
ISBN 0 7176 0466 7

HSG31
Pie and tart machines
ISBN 0 11 883891 1

HSG42
Safety in the use of metal
cutting guillotines and
shears
ISBN 0 11 885455 0

HSG45
Safety in meat preparation:
guidance for butchers
ISBN 0 7176 0781 X

HSG55
Health and safety in
kitchens and food
preparation areas
ISBN 0 11 885427 5

HSG89
Safeguarding agricultural
machinery: advice for
designers, manufacturers,
suppliers and users
ISBN 0 7176 2400 5

L22
Safe use of work equipment.
Provision and Use of Work
Equipment Regulations
1998. Approved Code of
Practice and guidance
ISBN 0 7176 1626 6

L112
Safe use of power presses.
Provision and Use of Work
Equipment Regulations 1998
as applied to power presses.
Approved Code of Practice
and guidance
ISBN 0 7176 1627 4

L113
Safe use of lifting equipment.
Lifting Operations and
Lifting Equipment Regula-
tions 1998. Approved Code of
Practice and guidance
ISBN 0 7176 1628 2

L114
Safe use of woodworking
machinery. Provision and
Use of Work Equipment
Regulations 1998 as applied
to woodworking machinery.
Approved Code of Practice
and guidance
ISBN 0 7176 1630 4

PM35
Safety in the use of reversing
dough brakes
ISBN 0 11 883576 9

PM65
Worker protection at
crocodile (alligator) shears
ISBN 0 11 883935 7

PM66
Scrap baling machines
ISBN 0 7176 1264 3

PM83
Drilling machines: guarding
of spindles and attachments
ISBN 0 7176 1546 4

* INDG229
Using work equipment
safely

* INDG270
Supplying new machinery

* INDG271
Buying new machinery

* INDG290
Simple guide to the Lifting
Operations and Lifting
Equipment Regulations 1998

* INDG291
Simple guide to the
Provision and Use of Work
Equipment Regulations
1998

* INDG297
Safety in gas welding,
cutting and similar
processes

5 Gas- and oil-fired equipment

COP20
Standards of
training in safe
gas installation
ISBN 0 7176 0603 1

HSG16
Evaporating and other ovens
ISBN 0 11 883433 9

L56
Safety in the installation
and use of gas systems and

appliances. Gas Safety
(Installation and Use)
Regulations 1998. Approved
Code of Practice and
guidance
ISBN 0 7176 1635 5

* INDG238
Gas appliances - Get them
checked, keep them safe

* INDG285
Landlords. A guide to landlords'
duties: Gas Safety (Installation
and Use) Regulations 1998

6 Plant and equipment maintenance

HSG62
Health and safety in tyre
and exhaust premises
ISBN 0 11 885594 8

HSG67
Health and safety in motor
vehicle repair
ISBN 0 7176 0483 7

L24
Workplace health, safety
and welfare. Workplace
(Health, Safety and Welfare)
Regulations 1992.
Approved Code of Practice
and guidance
ISBN 0 7176 0413 6

PM38
Selection and use of
electric handlamps
ISBN 0 11 886360 6

PM55
Safe working with
overhead travelling cranes
ISBN 0 11 883524 6

7 Pressurised plant and systems

COP37
Safety of pressure
systems. Pressure
Systems and
Transportable Gas
Containers Regulations 1989
ISBN 0 7176 0477 2

GS4
Safety in pressure testing
ISBN 0 7176 1629 0

HSG39
Compressed air safety
ISBN 0 7176 1531 6

PM5
Automatically controlled
steam and hot water boilers
ISBN 0 7176 1028 4

PM29
Electrical hazards from
steam/water pressure
cleaners etc
ISBN 0 7176 0813 1

PM60
Steam boiler blowdown
systems
ISBN 0 7176 1533 2

Pressure Systems and
Transportable Gas
Containers Regulations
1989: an open learning
course
ISBN 0 7176 0687 2

* INDG68
Do you use a steam/water
pressure cleaner?

* INDG261
Pressure systems: safety
and you

PLANT AND MACHINERY

8 Handling and transporting

COP
Safety of loads on vehicles
(Dept of Transport)
(Available from the
Stationery Office)
ISBN 0 11 550666 7

HSG6
Safety in working
with lift trucks
ISBN 0 7176 1440 9

HSG60
Work related upper limb
disorders: A guide to
prevention
ISBN 0 7176 0475 6

HSG115
Manual handling - solutions
you can handle
ISBN 0 7176 0693 7

HSG119
Manual handling in drinks
delivery
ISBN 0 7176 0731 3

HSG121
A pain in your workplace:
Ergonomic problems and
solutions
ISBN 0 7176 0668 6

HSG136
Workplace transport
safety - guidance for
employers
ISBN 0 7176 0935 9

L23
Manual handling. Manual
Handling Operations
Regulations 1992.
Guidance on Regulations
ISBN 0 7176 2415 3

L24
Workplace health, safety
and welfare. Workplace
(Health, Safety and
Welfare) Regulations 1992.
Approved Code of Practice
and guidance
ISBN 0 7176 0413 6

PM15
Safety in the use of timber
pallets
ISBN 0 7176 0714 3

Safe handling of bales
ISBN 0 7176 0692 9

* INDG125(rev1)
Handling and stacking
bales in agriculture

* INDG143
Getting to grips with
manual handling: A short
guide for employers

* INDG145
Watch your back:
avoiding back strain in
timber handling and
chainsaw work

* INDG148
Reversing vehicles

* INDG171
Upper limb disorders:
Assessing the risks

* INDG199
Managing vehicle safety at
the workplace: A short
guide for employers

9 Noise

HSG138
Sound solutions: Techniques
to reduce noise at work
ISBN 0 7176 0791 7

L108
Reducing noise at work:
guidance on the Noise at
Work Regulations 1989
ISBN 0 7176 1511 1

PM56
Noise from pneumatic
systems
ISBN 0 11 883529 7

* INDG75 (Rev)
Introducing the Noise at
Work Regulations

* INDG99
Noise at work: Advice for
employees

* INDG127 (Rev)
Noise in construction

* INDG193
Health surveillance in noisy
industries: Advice for
employers

* INDG263
Keep the noise down:
advice for purchasers of
workplace machinery

* INDG298
Ear protection: employers'
duties explained

10 Vibration

HSG88
Hand-arm vibration
ISBN 0 7176 0743 7

HSG170
Vibration solutions:
Practical ways to reduce
the risks of hand-arm
vibration injury
ISBN 0 7176 0954 5

* INDG126(rev1)
Health risks from hand-arm
vibration: Advice for
employees and the self-
employed

* INDG175(rev1)
Hand-arm vibration: Advice
for employers

* INDG242
In the driving seat: advice
to employers on reducing
back pain in drivers and
machinery operators

11 Radiations

COP23
Exposure to radon:
The Ionising Radiations
Regulations 1985
ISBN 0 11 883978 0

HSG95
Radiation safety of lasers
used for display purposes
ISBN 0 7176 0691 0

L7
Dose limitation - restriction
of exposure.
Approved Code of Practice
ISBN 0 11 885605 7

L49
Protection of outside
workers against ionising
radiations.
ISBN 0 7176 0681 3

L58
The protection of persons
against ionising radiation
arising from any work
activity
ISBN 0 7176 0508 6

Printing IAC
Safety in the use of inks,
varnishes and lacquers
cured by ultraviolet light or
electron beam techniques
ISBN 0 11 882045 1

Radiation safety for site
radiography (Engineering
Construction Industry
Association, Broadway
House, Tothill Street,
London SW1H 9NQ)

* INDG147(rev1)
Keep your top on

* INDG209
Controlling health risks
from the use of UV tanning
equipment

* INDG224
Controlling the radiation
safety of display laser
installations

12 Electricity

GS6
Avoidance of danger from overhead electrical lines
ISBN 0 7176 1348 8

GS50
Electrical safety at places of entertainment
ISBN 0 7176 1387 9

HSG85
Electricity at work: safe working practices
ISBN 0 7176 0442 X

HSG107
Maintaining portable and transportable electrical equipment
ISBN 0 7176 0715 1

HSG118
Electrical safety in arc welding
ISBN 0 7176 0704 6

HSG141
Electrical safety on construction sites
ISBN 0 7176 1000 4

HSR25
Memorandum of guidance on the Electricity at Work Regulations 1989
ISBN 0 11 883963 2

PM29
Electrical hazards from steam/water pressure cleaners etc
ISBN 0 7176 0813 1

PM38
Selection and use of electric handlamps
ISBN 0 11 886360 6

Video - Live wires
available on sale or hire from HSE Videos, PO Box 35, Wetherby, West Yorkshire LS23 7EX
Tel: 0845 741 9411

* INDG139
Electric storage batteries: safe charging and use

* INDG231
Electrical safety and you

* INDG236
Maintaining portable electrical equipment in offices and other low-risk environments

* INDG237
Maintaining portable electrical equipment in hotels and tourist accommodation

SUBSTANCES

13 Harmful substances

COP2
Control of lead at work
ISBN 0 7176 1506 5

EH10
Asbestos - exposure limits and measurement of air-borne dust concentrations
ISBN 0 7176 0907 3

EH40
Occupational exposure limits (updated annually)
ISBN 0 7176 1660 6

EH44
Dust: general principles of protection
ISBN 0 7176 1435 2

EH47
Provision, use and maintenance of hygiene facilities for work with asbestos insulation and coatings
ISBN 0 11 885567 0

GS46
In situ timber treatment using timber preservatives
ISBN 0 11 885413 5

HSG54
Maintenance, examination and testing of local exhaust ventilation
ISBN 0 7176 1485 9

HSG70
The control of legionellosis including legionnaires' disease
ISBN 0 7176 0451 9

HSG77
COSHH and peripatetic workers
ISBN 0 11 885733 9

HSG97
A step by step guide to COSHH assessment
ISBN 0 7176 1446 8

HSG110
Seven steps to successful substitution of hazardous substances
ISBN 0 7176 0695 3

HSG117
Making sense of NONS
ISBN 0 7176 0774 7

HSG126
CHIP 2 for everyone. Chemicals (Hazard Information and Packaging for Supply) Regulations 1994
ISBN 0 7176 0857 3

HSG173
Monitoring strategies for toxic substances
ISBN 0 7176 1411 5

HSG187
Control of diesel engine exhaust emissions in the workplace
ISBN 0 7176 1662 2

HSG188
Health risks management: a guide to working with solvents
ISBN 0 7176 1664 9

HSG189/1
Controlled asbestos stripping techniques for work requiring a licence
ISBN 0 7176 1666 5

HSG189/2
Working with asbestos cement
ISBN 0 7176 1667 3

HSG193
COSHH essentials: easy steps to control chemicals
ISBN 0 7176 2421 8

L5
General COSHH ACOP (Control of substances hazardous to health) and Carcinogens ACOP (Control of carcinogenic substances) and Biological agents ACOP (Control of biological agents). Control of Substances Hazardous to Health Regulations 1999. Approved Codes of Practice
ISBN 0 7176 1670 3

L8
The prevention or control of legionellosis (including legionnaires' disease). Approved Code of Practice
ISBN 0 7176 0732 1

L9
The safe use of pesticides for non-agricultural purposes. Approved Code of Practice
ISBN 0 7176 0542 6

L11
A guide to the Asbestos (Licensing) Regulations 1983 as amended
ISBN 0 7176 2435 8

L27
The control of asbestos at work. Control of Asbestos at Work Regulations 1987. Approved Code of Practice
ISBN 0 7176 1673 8

L28
Work with asbestos insulation, asbestos coating and asbestos insulating board. Control of Asbestos at Work Regulations 1987. Approved Code of Practice
ISBN 0 7176 1674 6

L55
Preventing asthma at work
ISBN 0 7176 0661 9

L62
Safety data sheets for substances and preparations dangerous for supply (CHIP 2). Approved Code of Practice
ISBN 0 7176 0859 X

L88
Approved methods for the classification and packaging of dangerous goods for carriage by road and rail
ISBN 0 7176 1221 X

L100
Approved guide to the classification and labelling of substances and preparations dangerous for supply (CHIP 97)
ISBN 0 7176 1366 6

L115
Approved supply list (CHIP 96, 97 and 98). 4th edition
ISBN 0 7176 1641 X

Approved supply list (supplement to 4th edition)
ISBN 0 7176 1683 5

Pesticides: code of practice for the safe use of pesticides on farms and holdings (HSC/MAFF) (Available from the Stationery Office)
ISBN 0 11 242892 4

13 Harmful substances (continued)

* INDG95 (Rev)
Respiratory sensitisers
and COSHH

* INDG136(rev1)
COSHH: A brief guide to
the Regulations

* INDG140
Grain dust in non-
agricultural workplaces

* INDG165
Health surveillance
programmes for employees
exposed to metalworking
fluids

* INDG167
Health risks from
metalworking fluids

* INDG168
Management of metalworking
fluids

* INDG169
Metalworking fluids and you

* INDG172
Pocket card
Breathe freely

* INDG181(rev1)
The complete idiot's guide
to CHIP

* INDG182
Why do I need a safety data
sheet?

* INDG186
Read the label: How to find
out if chemicals are
dangerous

* INDG223(rev1)
Managing asbestos in
workplace buildings

* INDG255(rev1)
Asbestos dust kills. Keep
your mask on

* INDG257
Pesticides: use them safely

* INDG273
Working safely with solvents:
a guide to safe working
practices

* INDG286
Diesel engine exhaust
emissions

* INDG289
Working with asbestos in
buildings

* MISC155
Substitutes for chrysotile
(white asbestos)

* MSA1(rev1)
Lead and you

14 Flammable and explosive substances

HSG51
The storage of
flammable liquids
in containers
ISBN 0 7176 1471 9

HSG71
Chemical warehousing:
storage of packaged
dangerous substances
ISBN 0 7176 1484 0

HSG92
Safe use and storage of
cellular plastics
ISBN 0 7176 1115 9

HSG103
Safe handling of
combustible dusts
ISBN 0 7176 0725 9

HSG114
Conditions for the
authorisation of explosives
in Great Britain
ISBN 0 7176 0717 8

HSG117
Making sense of NONS
ISBN 0 7176 0774 7

HSG126
CHIP 2 for everyone.
Chemicals (Hazard,
Information and Packaging
for Supply) Regulations
1994
ISBN 0 7176 0857 3

HSG131
Energetic and
spontaneously combustible
substances
ISBN 0 7176 0893 X

HSG135
Storage and handling of
industrial nitrocellulose
ISBN 0 7176 0694 5

HSG140
The safe use and handling of
flammable liquids
ISBN 0 7176 0967 7

HSG146
Dispensing petrol
ISBN 0 7176 1048 9

HSG150
Health and safety in
construction
ISBN 0 7176 1143 4

HSG176
The storage of flammable
liquids in tanks
ISBN 0 7176 1470 0

HSG178
The spraying of flammable
liquids
ISBN 0 7176 1483 2

HSR17
Guide to the Classification
and Labelling of Explosives
Regulations 1983
ISBN 0 11 883706 0

Carriage of dangerous goods explained

Part 1	HSG160 (plus supplement) ISBN 0 7176 1255 4	
Part 2	HSG161 (plus supplement) ISBN 0 7176 1253 8	
Part 3	HSG163 (plus supplement) ISBN 0 7176 1256 2	
Part 4	HSG162 (plus supplement) ISBN 0 7176 1675 4	
Part 5	HSG164 (plus supplement) ISBN 0 7176 1257 0	

L10
A guide to the Control of
Explosives Regulations 1991
ISBN 0 11 885670 7

L13
A guide to the Packaging of
Explosives for Carriage
Regulations 1991
ISBN 0 11 885728 2

L62
Safety data sheets for
substances and preparations
dangerous for supply
(CHIP 2). Approved Code
of Practice
ISBN 0 7176 0859 X

L88
Approved methods for the
classification and packaging
of dangerous goods for
carriage by road and rail
ISBN 0 7176 1221 X

L89
Approved vehicle
requirements
ISBN 0 7176 1680 0

L90
Approved carriage list
ISBN 0 7176 1681 9

L92
Approved requirements for
the construction of vehicles
intended for the carriage of
explosives by road
ISBN 0 7176 1679 7

L100
Approved guide to the
classification and labelling
of substances and
preparations dangerous for
supply (CHIP 97)
ISBN 0 7176 1366 6

List of classified and
authorised explosives 1994
(LOCAE)
ISBN 0 7176 0772 0

L115
Approved supply list (CHIP
96, 97 and 98). 4th edition
ISBN 0 7176 1641 X

Approved supply list
(supplement to 4th edition)
ISBN 0 7176 1683 5

* CHIS4
Use of LPG in small bulk tanks

* CHIS5
Small-scale use of LPG in
cylinders

* HSE8(rev2)
Take care with oxygen

* INDG115
New explosives controls: The
Control of Explosives
Regulations 1991

* INDG181(rev1)
The complete idiot's guide
to CHIP

* INDG182
Why do I need a safety data
sheet?

* INDG186
Read the label: How to find out
if chemicals are dangerous

* INDG216
Petrol filling stations: Control
and safety guidance for
employees

* INDG227
Safe working with flammable
substances

* INDG234(rev)
Are you involved in the
carriage of dangerous goods by
road or rail?

* INDG273
Working safely with solvents:
a guide to safe working
practices

PROCEDURES

15 Safe systems

CS15
Cleaning and gas freeing of tanks containing flammable residues
ISBN 0 7176 1365 8

HSG85
Electricity at work: safe working practices
ISBN 0 7176 0442 X

L101
Safe work in confined spaces
ISBN 0 7176 1405 0

* INDG73(rev1)
Working alone in safety

* INDG98(Rev)
Chemical manufacturing: permit to work systems

* INDG258
Safe work in confined spaces

16 Accidents and emergencies

Form F2508/ F2508A
Report of injury or dangerous occurrence and Report of a case of disease
ISBN 0 7176 1078 0

HSR29
Notification and marking of

sites. The Dangerous Substances (Notification and Marking of Sites) Regulations 1990
ISBN 0 11 885435 6

L73
Guide to the Reporting of Injuries, Diseases and Dangerous Occurrences

Regulations 1995
ISBN 0 7176 2431 5

L74
First aid at work. Health and Safety (First Aid) Regulations 1981
ISBN 0 7176 1050 0

* HSE31(rev1)
RIDDOR explained

* INDG214
First aid at work: Your questions answered

* INDG215(rev2)
Basic advice on first aid at work

* INDG246
Chemicals - prepared for emergency!

PEOPLE

17 Health care

HSG61
Surveillance of people exposed to health risks at work
ISBN 0 7176 0525 6

HSG100
Prevention of violence to staff in banks and building societies
ISBN 0 7176 0683 X

HSG116
Stress at work: A guide for employers
ISBN 0 7176 0733 X

HSG122
New and expectant mothers at work: A guide for employers
ISBN 0 7176 0826 3

HSG133
Preventing violence to retail staff
ISBN 0 7176 0891 3

HSG167
Biological monitoring in the workplace: a guide to its practical application to chemical exposure
ISBN 0 7176 1279 1

L21
Management of Health and Safety at Work Regulations 1992. Approved Code of Practice ISBN 0 7176 0412 8

MS24
Health surveillance of occupational skin disease
ISBN 0 7176 1545 6

Health surveillance under COSHH: Guidance for employers
ISBN 0 7176 0491 8

Infections in the workplace to new and expectant mothers
ISBN 0 7176 1360 7

Violence and aggression to staff in the health services
ISBN 0 7176 1466 2

* INDG62
Protecting your health at work

* INDG63(Rev)
Passive smoking at work

* INDG69(Rev)
Violence at work

* INDG91(rev2)
Drug misuse at work

* INDG116
What your doctor needs to know

* INDG240
Don't mix it: A guide for employers on alcohol at work

* INDG245
Biological monitoring for chemicals in the workplace: information for employees on its application to chemical exposure

* INDG281
Help on work-related stress: a short guide

18 Personal protective equipment

HSG53
The selection, use and maintenance of respiratory protective equipment: A practical guide for users
ISBN 0 7176 1537 5

L25
Personal Protective Equipment at Work Regulations 1992. Guidance on Regulations
ISBN 0 7176 0415 2

L102
Construction (Head Protection) Regulations 1989 - Guidance on Regulations
ISBN 0 7176 1478 6

* INDG174
A short guide to the

Personal Protective Equipment at Work Regulations 1992

* INDG288
Selection of suitable respiratory protective equipment for work with asbestos

19 Selection and training

GS48
Training and standards of competence for people working with chainsaws
ISBN 0 7176 1403 4

HSG83
Training woodworking machinists
ISBN 0 11 886316 9

HSG165
Young people at work: A guide for employers
ISBN 0 7176 1285 6

L21
Management of Health and Safety at Work Regulations 1992. Approved Code of Practice
ISBN 0 7176 0412 8

L117
Rider-operated lift trucks: operator training
ISBN 0 7176 2455 2

* INDG213
Five steps to information, instruction and training

* INDG235
A guide to information, instruction and training: Common provisions in health and safety law

* TOP06
Supervising for safety in woodworking

How to obtain publications

HSE publications

HSE publications are available by mail order from:

HSE Books,
PO Box 1999,
Sudbury,
Suffolk CO10 2WA
Tel: 01787 881165
Fax: 01787 313995
Website: www.hsebooks.co.uk

HSE priced publications are also available from all good booksellers.

For other enquiries ring HSE's Infoline,
Tel: 08701 545500 or write to
HSE's Infoline,
HSE Information Centre,
Broad Lane, Sheffield S3 7HQ
Website: www.hse.gov.uk

The HSC Newsletter is available on subscription from:

HSE Books,
Subscriptions Department,
PO Box 1999,
Sudbury,
Suffolk CO10 2WA
Tel: 01787 881165
Fax: 01787 313995

Other publications

The Stationery Office (formerly HMSO) publications are available from:

The Publications Centre,
PO Box 276,
London SW8 5DT
Tel: 0870 600 5522
Fax: 020 7873 8200

British Standards are available from:

BSI Sales and Customer Services
389 Chiswick High Road
Chiswick
London W4 4AL
Tel: 020 8996 7000
Fax: 020 8996 7001

Other sources of information

British Safety Council,
National Safety Centre,
70 Chancellors Road,
London W6 9RS
Tel: 020 8741 1231

Royal Society for the
Prevention of Accidents,
Edgbaston Park,
353 Bristol Road,
Birmingham B5 7ST
Tel: 0121 248 2000

Trades Union Congress,
Congress House,
Great Russell Street,
London WC1B 3LS
Tel: 020 7636 4030

Confederation of
British Industry,
Centre Point,
103 New Oxford Street
London WC1A 1DA
Tel: 020 7379 7400

Printed and published by the Health and Safety Executive   11/00   C250